Advanced Statistical Modeling, Forecasting, and Fault Detection in Renewable Energy Systems

Edited by Fouzi Harrou and Ying Sun

Published in London, United Kingdom

IntechOpen

Supporting open minds since 2005

Advanced Statistical Modeling, Forecasting, and Fault Detection in Renewable Energy Systems
http://dx.doi.org/10.5772/intechopen.85999
Edited by Fouzi Harrou and Ying Sun

Contributors
Pushpavalli Murugan, Nassim Sabri, Abdelhalim Tlemçani, Aissa Chouder, Harrouz Abdelkader, Tahiri Fadila, Boussaid Ibrahim, Bekraoui Fatiha, Sid-Ali Blaifi, Bilal Taghezouit, Youssef McHaouar, Abdelmajid Abouloifa, Ibtissam Lachkar, Mohammed Fettach, Nouar Aoun, Fouzi Harrou, Hong Son Hoang, Rémy Baraille, Haoran Zhang, Yamin Yan, Jianqin Zheng, Yongtu Liang, Bharathi M.L

First published in London, United Kingdom, 2020 by IntechOpen
IntechOpen is the global imprint of INTECHOPEN LIMITED, registered in England and Wales, registration number: 11086078, 7th floor, 10 Lower Thames Street, London, EC3R 6AF, United Kingdom
Printed in Croatia

British Library Cataloguing-in-Publication Data
A catalogue record for this book is available from the British Library

Additional hard and PDF copies can be obtained from orders@intechopen.com

Advanced Statistical Modeling, Forecasting, and Fault Detection in Renewable Energy Systems
Edited by Fouzi Harrou and Ying Sun
p. cm.
Print ISBN 978-1-83880-091-8
Online ISBN 978-1-83880-092-5
eBook (PDF) ISBN 978-1-83880-546-3

An electronic version of this book is freely available, thanks to the support of libraries working with Knowledge Unlatched. KU is a collaborative initiative designed to make high quality books Open Access for the public good. More information about the initiative and links to the Open Access version can be found at www.knowledgeunlatched.org

For EU product safety concerns:
IN TECH d.o.o., Prolaz Marije Krucifikse Kozulić 3, 51000 Rijeka, Croatia, info@intechopen.com or visit our website at intechopen.com.

Meet the editors

Fouzi Harrou received his M.Sc. degree in telecommunications and networking from the University of Paris VI, France, and his Ph.D. degree in systems optimization and security from the University of Technology of Troyes (UTT), France. He was an Assistant Professor with UTT for one year and with the Institute of Automotive and Transport Engineering, Nevers, France, for one year. He was also a Postdoctoral Research Associate with the Systems Modeling and Dependability Laboratory, UTT, for one year. He was a Research Scientist with the Chemical Engineering Department, Texas A&M University, Doha, Qatar, for three years. He is currently a Research Scientist with the Division of Computer, Electrical and Mathematical Sciences and Engineering, King Abdullah University of Science and Technology. He is the author of more than 100 refereed journals, conference publications, and book chapters. His current research interests include statistical decision theory and its applications, fault detection and diagnosis, and deep learning.

Ying Sun received her Ph.D. degree in statistics from Texas A&M, in 2011. She held a two-year postdoctoral research position at the Statistical and Applied Mathematical Sciences Institute and the University of Chicago. She was an Assistant Professor with Ohio State University for a year before joining KAUST in 2014. At KAUST, she established and leads the Environmental Statistics research group, which works on developing statistical models and methods for complex data to address important environmental problems. She has made original contributions to environmental statistics, in particular in the areas of spatiotemporal statistics, functional data analysis, visualization, computational statistics, with an exceptionally broad array of applications. She received two prestigious awards: The Early Investigator Award in Environmental Statistics presented by the American Statistical Association and the Abdel El-Shaarawi Young Research Award from The International Environmetrics Society.

Contents

Preface

Renewable energy is a key challenging problem that is increasingly gaining attention worldwide. Renewable energy sources, such as solar and wind, are promising alternatives to conventional fossil fuels because they are clean, sustainable, safe, and environmentally friendly with zero CO2 emissions. For instance, it has been shown that 100 GW of photovoltaic (PV)-generated power in Europe in 2012 kept more than 53 million tons of CO2 from being emitted into the atmosphere. On the other hand, the global capacity of wind power has reached 539 GW in 2017, with an increase of 52.5 GW compared to 2016. According to the World Wind Energy Association, the overall capacity of all wind turbines installed by the end of 2018 attained 600 GW. Furthermore, solar PV and wind energy, which are sustainable and economically competitive renewable energy sources, increase a countries' energy security by reducing their dependence on fossil fuels. Hence, the requirement to operate these sources becomes a mandatory obligation to reduce environment pollution and participate in sustainable energy development.

However, the main crucial and challenging issue in solar energy production is the volatility of intermittent power generation due mainly to weather conditions. In particular, a variation of the temperature and irradiance can have a profound impact on the quality of electric power production. A drop of more than 20% of power PV production can be observed in real PV energy plants. This fact usually limits the integration of PV systems into the power grid. Hence, accurately forecasting the power output of PV modules in the short-term is of great importance for daily/hourly efficient management of power grid production, delivery, and storage, as well as for decision-making on the energy market. Such management aims to take suitable control actions to balance the electricity supply and the large electricity demand, as well as to meet an electricity generation performance complying with power quality requirements and standards.

Despite the progress in renewable energy systems to ensure their optimal operation, in practice several factors can significantly affect their performance by decreasing their efficiency. Indeed, PV systems and wind turbines are frequently exposed to different sources of faults and failures that affect the power generated by these systems. For example, in a PV system, many potential external interferences or faults can cause significant losses in power production, such as dust accumulation on the PV modules and shading. These faults could considerably reduce both their production efficiencies and lifespan. Therefore, it is crucial to detect and identify possible faults or failures in renewable energy systems (i.e., PV systems and wind turbines) as early as possible. Hence, accurate and prompt fault detection and diagnosis are required to improve renewable energy systems efficiencies, avoid the high cost of maintenance, and reduce risks of fire hazards, which could affect both personnel and installed equipment.

The main objective of this book is to study and develop efficient statistical modeling, forecasting, and fault detection techniques for renewable energy systems. To this end, different methods are presented to model and forecast power production using a model-based or data-based framework. Moreover, innovative techniques are proposed for fault isolation and control in renewable energy systems. The book is divided into ten chapters. These chapters are regrouped into three parts: forecasting, modeling, and fault detection and control.

We would like to thank Marijana Francetic, Author Service Manager, for her continuous assistance during the preparation of this book. We would like to gratefully acknowledge the financial support by funding from King Abdullah University of Science and Technology (KAUST), Office of Sponsored Research (OSR) under Award No: OSR-2019-CRG7-3800.

Fouzi Harrou and Ying Sun
King Abdullah University of Science and Technology (KAUST) Computer, Electrical and Mathematical Sciences and Engineering (CEMSE) Division, Thuwal, Saudi Arabia

Section 1

Forecasting in Renewable Energy Systems

Chapter 1

Forecasting of Photovoltaic Solar Power Production Using LSTM Approach

Fouzi Harrou, Farid Kadri and Ying Sun

Abstract

Solar-based energy is becoming one of the most promising sources for producing power for residential, commercial, and industrial applications. Energy production based on solar photovoltaic (PV) systems has gained much attention from researchers and practitioners recently due to its desirable characteristics. However, the main difficulty in solar energy production is the volatility intermittent of photovoltaic system power generation, which is mainly due to weather conditions. For the large-scale solar farms, the power imbalance of the photovoltaic system may cause a significant loss in their economical profit. Accurate forecasting of the power output of PV systems in a short term is of great importance for daily/hourly efficient management of power grid production, delivery, and storage, as well as for decision-making on the energy market. The aim of this chapter is to provide reliable short-term forecasting of power generation of PV solar systems. Specifically, this chapter presents a long short-term memory (LSTM)-based deep learning approach for forecasting power generation of a PV system. This is motivated by the desirable features of LSTM to describe dependencies in time series data. The performance of the algorithm is evaluated using data from a 9 MWp grid-connected plant. Results show promising power forecasting results of LSTM.

Keywords: forecasting, deep learning, LSTM, solar power production

1. Introduction

Solar energy becomes one of the most promising sources for generating power for residential, commercial, and industrial applications [1, 2]. Solar photovoltaic (PV) systems use PV cells that convert solar irradiation into electric power. Renewable energy sources, in particular photovoltaic (PV) energy, has been progressively increased in recent years because of its advantages of being plentiful, inexhaustible, clean energy and environmentally friendly [3–5]. As one of the most popular renewable energy sources, solar energy has the advantages of abundant resources, no pollution, free use, and no transportation [6–8]. This greatly accelerated the installation of solar photovoltaic (PV) systems around the world.

Reliable and precise forecasting plays an important role in enhancing power plant generation based on renewable energy sources such as water, wind, and sun [9]. One of the most sustainable and competitive renewable energy sources is solar photovoltaic (PV) energy which is becoming nowadays more attracting than ever before [3]. The main crucial and challenging issue in solar energy production is the volatility intermittent of PV system power generation due to mainly to weather conditions. In particular, a variation of the temperature and irradiance can have a profound impact on the quality of electric power production. A drop of more than 20% of power PV production can be observed in real PV energy plants. This fact usually limits the integration of PV systems into the power grid. Hence, accurately forecasting the power output of PV modules in a short-term is of great importance for daily/hourly efficient management of power grid production, delivery, and storage, as well as for decision-making on the energy market [10].

Precise forecasting of solar energy is important for photovoltaic (PV) based energy plants to facilitate early participation in energy auction markets and efficient resource planning [11]. Numerous methods have been reported in the literature for PV solar power forecasting. These methods can be classified into four classes: (i) statistical approaches based on data-driven formulation to forecast solar time series by using historical measured data, (ii) machine learning techniques, in particular, deep learning approaches based artificial neural network, (iii) physical models based on numerical weather prediction and satellite images, and (iv) hybrid approaches which are the combination of the above methods. In [12], a combined approach merging seasonal autoregressive integrated moving average (SARIMA), random vector functional link neural network hybrid model and discrete wavelet transform has been introduced for forecasting short-term solar PV power production. It has been shown that the combined models provide improved forecasting results compared to individuals ones. In [13], Gradient boosted regression trees approach has been used to predict solar power generation for 1–6 h ahead. It has been that this approach outperforms the simpler autoregressive models. In [14], a model combining seasonal decomposition and least-square support vector regression has been designed to forecast power output. This approach demonstrated good forecasting capacity compared to the autoregressive integrated moving average (ARIMA), SARIMA, and generalized regression neural network. In [15], a multivariate ensemble forecast framework integrating ensemble framework with neural predictors and Bayesian adaptive combination is proposed for forecasting PV output power.

Most conventional solar power forecasting approaches are limited in uncovering the correlation of the limited data but are not able to deep correlation and uncover implicit and relevant information. With the huge data from the modern power system, the use of conventional approaches is not suited for guaranteeing precise forecasting. Recently, deep learning (DL) approaches have emerged as powerful machine learning tools that enable complicated pattern recognition and regression analysis and prediction applications [16–18]. DL approaches are becoming increasingly popular due to their good capacity in describing dependencies in time series data. Deep Learning is the result of the concatenation of more layers into the neural network framework. Over the past few decades, many deep learning models have been proposed including Boltzmann machines, Deep Belief Networks (DBN) and Recurrent Neural Networks (RNNs) [19]. RNN is a type of neural networks that exploits the sequential nature of input data. RNNs are used to model time-dependent data, and they give good results in the time series data, which have proven successful in several applications domains [3, 20, 21]. Long Short-Term Memory Networks (LSTM) is a type of RNNs that is able to deal with remembering information for much longer periods of time [22]. It is also considered as one of the most used RNN models for time series data predictions, which is perfectly suited to PV solar power production forecasting problems. In this chapter, we applied the LSTM model to accurately forecast short-term photovoltaic solar power. The effectiveness of this approach is tested based on power output data collected from a 9 MWp grid-connected plant.

The next section introduces the core idea behind the LSTM model and how it can be designed and implemented. Then, Section 3 presents the results of solar photovoltaic power forecasting using the LSTM model. Lastly, conclusions are offered in Section 4.

2. Deep learning and forecasting of PV power production

Over the last decades, many studies have been dedicated to forecasting problems in several application domains. Recurrent Neural Networks (RNNs) have been successfully used in machine learning problems [23]. These models have been proposed to address time-dependent learning problems [22]. **Figure 1** shows the basic concept of RNNs; a chunk of a neural network, A, looks at some input x_t and outputs a value h_t. It should be noted that RNNs are suited to learn and extract temporal information [24]. A general formula for RNN hidden state h t given an input sequence $\mathbf{x} = (\mathbf{x_1}, \mathbf{x_2}, \dots, \mathbf{x_t})$:

$$h_t = \begin{cases} 0, & t = 0 \\ \varphi(W_{x_t}, x_t), & otherwise \end{cases} \tag{1}$$

where φ is a non-linear function. The update of recurrent hidden state is realized as:

$$h_t = g(W_{x_t} + uh_{t-1}) \tag{2}$$

where g is a hyperbolic tangent function (*tanh*).

Generally, it is not easy to capture long term time dependencies in time series when using recurrent neural networks. To bypass this limitation, Long Short-Term Memory Networks (LSTM) models were designed. LSTM is an extended version of RNN that are effectively capable to handle time dependency in data [22]. These models are flexible and efficient to describe time-dependent data, and they demonstrated success in several applications. LSTM is one of the most used RNN models for time series data predictions, which is perfectly suited to the PV forecasting problems [22]. Next, we present a basic overview of LSTM and how it can be designed and implemented.

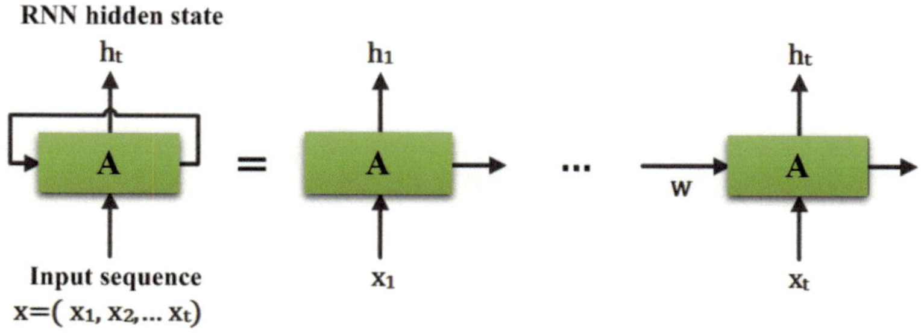

Figure 1.
Basic illustration of RNN.

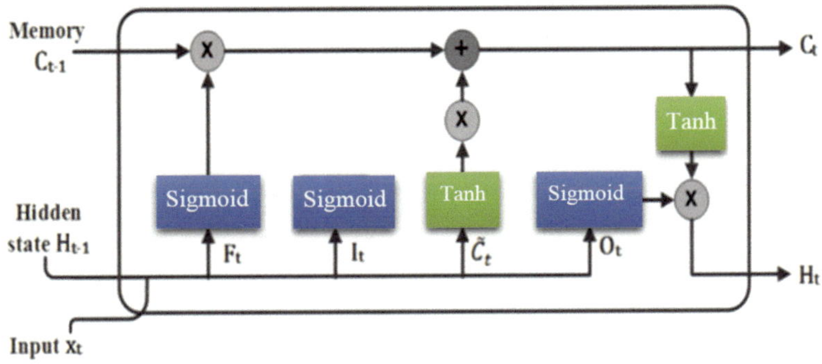

Figure 2.
Illustration of LSTM unit.

2.1 Long short-term memory (LSTM) models

The Long Short-Term Memory (LSTM) is a variant of the Recurrent Neural Networks (RNN) that is capable of learning long term dependencies. LSTM models were initially proposed by Hochreiter and Schmidhuber [4] and were improved and popularized by many other researchers [4–6, 9]. LSTM models have an excellent ability to memorize long-term dependencies, are developed to deal with the exploding and vanishing gradient problems that can be encountered when training traditional RNNs. Relative insensitivity to gap length is an advantage of LSTM models over ANNs models, hidden Markov models and other sequence learning methods in several application domains.

A common LSTM model is composed of cell blocks in place of standard neural network layers. These cells have various components called the input gate, the forget gate and the output gate. The cell remembers values over arbitrary time intervals and the three gates regulate the flow of information into and out of the cell [5]. **Figure 1** shows the basic structure of RNN-LSTM.

From **Figure 2**, the RNN-LSTM has two input features at each time, which include the current time step input X_t (input vector) and the hidden state of the previous time step H_{t-1} (previous input vector). The output is computed by the fully connected layer with its activation function (e.g., *tanh*, sigmoid, Softmax, and Adam). Therefore, the output of each gate can be obtained through logical operation and nonlinear transformation of input.

Let us denote the input time series as X_t, the number of hidden units as h, the hidden state of the last time step as H_{t-1}, and the output time series as H*t*.
The mathematical relationship between inputs and outputs of the RNN-LSTM can be described as follows.

$$I_t = \sigma(X_t W_{xi} + H_{t-1} W_{hi} + b_i) \tag{3}$$

$$F_t = \sigma(X_t W_{xf} + H_{t-1} W_{hf} + b_f) \tag{4}$$

$$O_t = \sigma(X_t W_{xo} + H_{t-1} W_{ho} + b_o) \tag{5}$$

$$\tilde{C}_t = \tanh(X_t W_{xc} + H_{t-1} W_{hc} + b_c) \tag{6}$$

$$C_t = F_t \, o \, C_{t-1} \, o \, \tilde{C}_t \tag{7}$$

$$H_t = O_t \, o \, \tanh(C_t) \tag{8}$$

where

- $\boldsymbol{I_t}$, $\boldsymbol{F_t}$, $\boldsymbol{O_t}$ are input gate, forget gate, and output gate respectively, $\boldsymbol{W_{xi}}$, $\boldsymbol{W_{xf}}$, $\boldsymbol{W_{xo}}$ and $\boldsymbol{W_{hi}}$, $\boldsymbol{W_{hf}}$, $\boldsymbol{W_{ho}}$ are weight parameters and $\boldsymbol{b_i}$, $\boldsymbol{b_f}$, $\boldsymbol{b_o}$ are bias parameters. All these gates have the same dimensions and the same equations just with different parameters. They are called gates because the activation function transforms the element values between ranges ([0, 1], [−1, 1]). The input gate defines how much of the newly computed state for the current input you want to let through. The forget gate defines how much of the previous state you want to let through. The output gate defines how much of the internal state you want to expose to the external network (higher layers and the next time step).

- $\tilde{C}_t$ is the candidate memory cells, W_{xc}, W_{hc} are weight parameters and b c is a bias parameter. LSTM model needs to compute the candidate memory cell $\tilde{C}_t$, its computation is similar to the three gates (input, forget and output gates), but using a *tanh* function as an activation function with a value range between [−1, 1].

- C_t is the memory cells, o is an operator that expresses element-wise multiplication. The computation of the current time steps memory cell $\mathbf{C_t}$ combines the information of the previous time step memory cells (C_{t-1}) and the current time step candidate memory cells ($\tilde{C}_t$), and controls the flow of information through forgetting gate and input.

- $\mathbf{H_t}$ is the hidden states, we can control the flow of information from memory cells to the hidden state $\mathbf{H_t}$ through the output gate. The *tanh* function ensures that the hidden state element value is between [−1, 1]. It should be noted that when the output gate is approximately 1, the memory cell information will be passed to the hidden state for use by the output layer; and when the output gate is approximately 0, the memory cell information is only retained by itself.

2.2 Proposed approach

The proposed approach in this chapter aims to forecast solar power production. This methodology is based on the LSTM deep-learning model. **Figure 3** summarizes the main steps of the proposed methodology. The proposed approach includes four key steps (**Figure 3**):

i. Collect the SCADA data from the PV system.

ii. Pre-process and clean data by removing outliers and imputing missing values.

iii. Normalize the original data.

iv. Train, validate and test the LSTM model. Various statistical indicators are used to quantify the accuracy of the developed model. Lastly, the designed LSTM model can be used for power production forecasting.

2.3 Metrics for evaluating the forecasting models

To assess the forecasting performance, numerous statistical indicators have been proposed in the literature including root mean square error (RMSE), mean absolute error (MAE), coefficient of determination (R^2), and mean absolute percentage

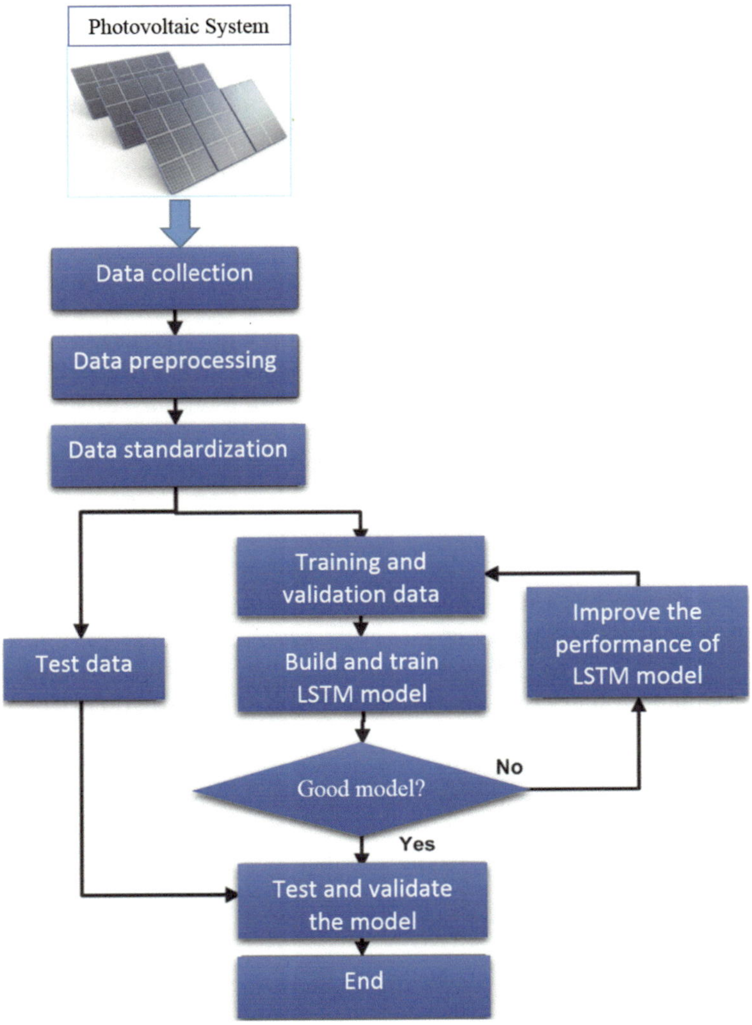

Figure 3.
Schematic block of the proposed forecasting method.

error (MAPE). In this study, we used R^2 and MAPE, which are frequently to evaluate the forecasting accuracy:

$$MAE = \frac{1}{n}\sum|\hat{x} - x| \tag{9}$$

$$RMSE = \sqrt{\frac{\sum(\hat{x} - x)^2}{n}} \tag{10}$$

$$R^2 = 1 - \frac{\sum_{i=1}^{N}(x_i - \hat{x}_i)^2}{\sum_{i=1}^{N}(x_i - \overline{x})^2} \tag{11}$$

where x are the measured values, $\hat{x}$ are the corresponding forecasted values by the LSTM model and n is the number of measurements.

2.4 Implementation steps

Essentially, the LSTM model can be designed and implemented in four main steps. At first, define the LSTM model and train it, then fit the LSTM model, and lastly, the trained LSTM model is used for forecasting. **Table 1** summarizes the main steps (partial codes) performed in designing the LSTM model.

2.5 Enhance LSTM models performance

The key factors impacting the accuracy of the LSTM model are not only the amount of training data but also the architecture of the network, hyper-parameters and the utilized optimizers. Accordingly, the performance of LSTMs can be enhanced by acting on the following elements.

- *Activation functions*: activation functions an important role in determining the final response of the neural network. Two families of functions are distinguished: linear and nonlinear functions. The output of the linear activation functions is linearly proportional to the inputs and is not limited between any ranges. They are more suited than a step function because they permit obtaining multiple outputs, not just binary output (i.e., yes and no). On the other hand, nonlinear activation functions are the most frequently utilized because they are flexible and permit obtaining nonlinear output and they are confined within a range. For instance, Sigmoid, Softmax, and Rectified Linear Unit (ReLU) activation functions permit rescaling the data to values in the interval $[0, 1]$, while Hyperbolic Tangent (tanh) activation functions rescale the data within $[-1, 1]$.

- *Optimizer*: in the training phase of the LSTM model, optimization algorithms are used for minimizing its error rate. The performance of an optimizer is generally characterized by convergence speed and generalization (the efficiency of the model on new datasets). The commonly used optimizers include Adaptive Moment Estimation (Adam) or Stochastic Gradient Descent (SGD) [25, 26].

- *Dropout*: it is a well-known stochastic regularization procedure applied to avoid overfitting and further enhance the prediction capacity of RNN models [27]. More details about dropout techniques can be found in [26–28].

- *Epochs and batches*: the number of epochs and batch are two important parameters when constructing deep learning models. It has been shown in the

Steps	Action
Step 1: Define LSTM network	*from keras.layers.recurrent import LSTM* *from keras.models import Sequential* *from keras.layers.core import Activation, Dense, Dropout* *model = Sequential()* *model.add(LSTM(units=nb_neural, return_sequences=True, input_shape=(Xtrain.shape[1], 1)))*
Step 2: Compile the LSTM network	*model.compile(loss="mse", optimizer="adam", metrics=[rmse, 'mae', Rsquare])*
Step 3: Fit the LSTM network	*history = model.fit(Xtrain, ytrain, batch_size=batch_size, epochs=num_epochs,* *validation_data=(Xval, yval), verbose=2)*
Step 4: Forecasting	*ypred = model.predict(Xtest)*

Table 1.
Partial codes used for building the LSTM network.

literature that good results can be achieved when using large epochs and small batch sizes.

- *Weight regularization*: another way to avoid overfitting and improves model performance is called weight regularization. This approach imposes constraints on the RNN weights within nodes to allow the network to maintain the weights small. Several penalizing or regularization approaches are commonly used in the literature based on L1 or L2 vector norm penalty.

3. Results and discussion

This study is based on real data collected from January 2018 to December 2018 every 15 min from a 9 MWp grid-connected plant. **Figure 4** shows the hourly distribution of PV power production day from January 2018 to December 2018.

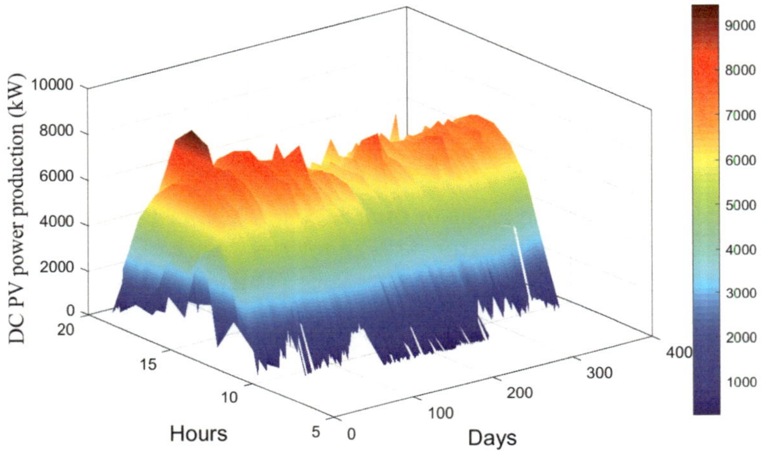

Figure 4.
PV power production per hour for each day from January 2018 to December 2018.

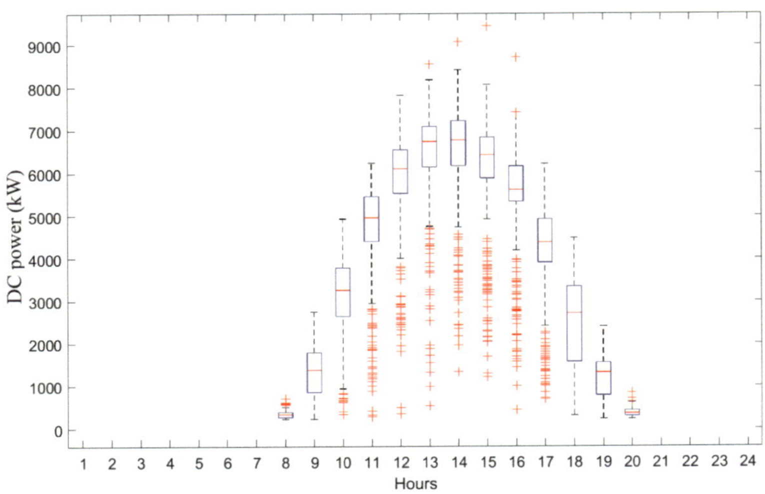

Figure 5.
Distribution of DC power output in the daytime.

From **Figure 4**, the solar PV power production reaches, every day, its maximum at the mid-day and falls to zero over the night.

For the data in **Figure 4**, the box plots showing the distribution of DC power generation in the daytime are displayed in **Figure 5**. One can see that the maximum power production is achieved around mid-day.

The monthly cumulative DC power generated by the inspected PV system from January 2018 to December 2018 is displayed in **Figure 6**. The highest and lowest monthly cumulative power are respectively achieved in March (6450.056 MW) and in October (4655.524 MW).

Figure 7 shows the monthly distribution of DC power production during the monitored period. **Figure 7** shows that the produced DC power is relatively high in January, February, and March. Also, it can be noticed that the production was relatively low from June to September (**Figure 7**).

To investigate the interactions between the DC power and meteorological factors (i.e., inclined irradiance 27, ambient temperature, and wind velocity) a Pearson correlation heatmap is displayed in **Figure 8**. From **Figure 8**, one can see that there

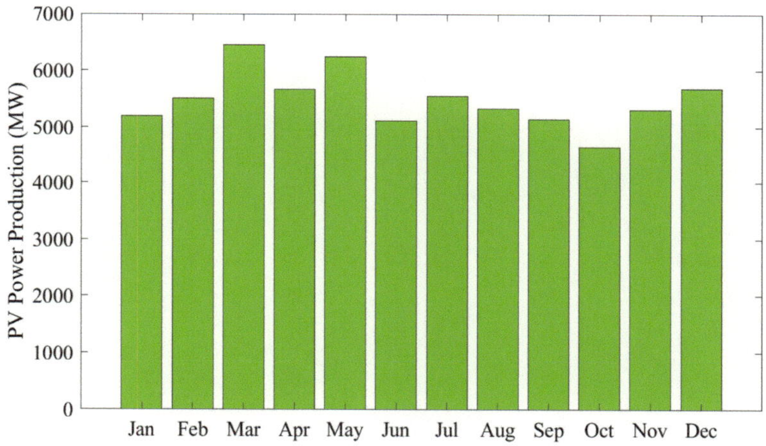

Figure 6.
Monthly total DC power produced from January 2018 to December 2018.

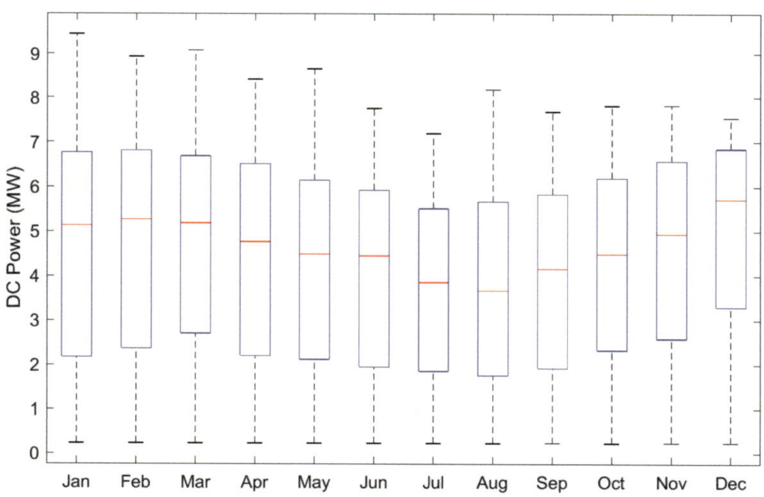

Figure 7.
Monthly distribution of DC power output.

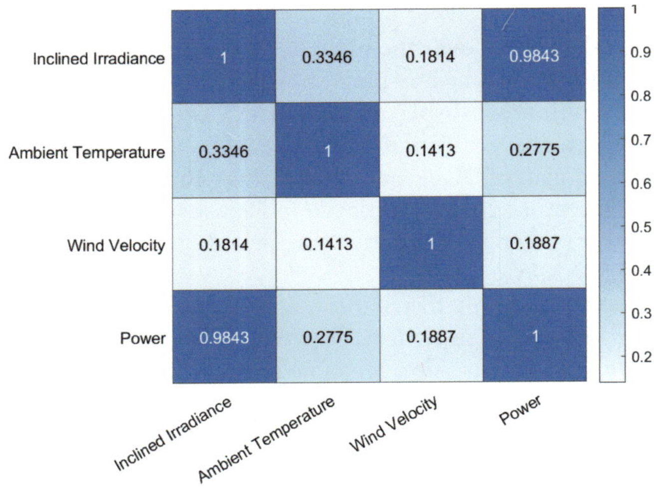

Figure 8.
Heatmap of the correlation matrix of data: inclined irradiance 27, ambient temperature, wind velocity, and power.

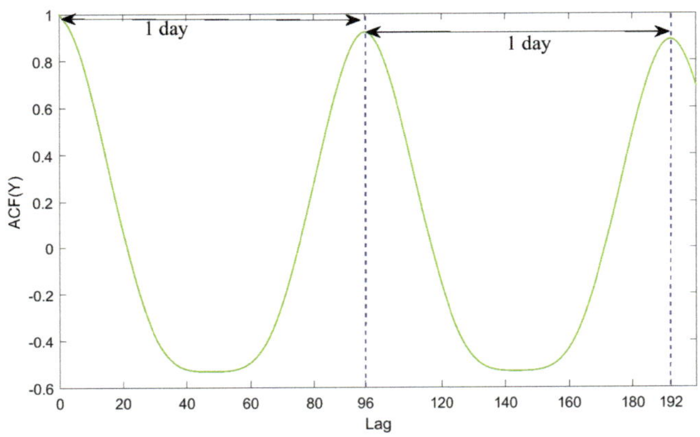

Figure 9.
ACF DC power measurements.

Models	Parameter	Value
	Number of LSTM layers	1
	learning rate	0.001
	Loss function	Cross-entropy
	Number of hidden units	200
	Optimizer	Adam
	Number of training epochs	100

Table 2.
Parameters in LSTM model.

is a high correlation between solar irradiance and power production. It should be noted that DC Power has a low correlation with wind velocity.

Figure 9 shows the autocorrelation function (ACF) plot of the data shown in power generation data. A seasonality of 24 h can be seen from the ACF plot of PV power data, the time difference between two maximum in the ACF (**Figure 9**). In particular, this seasonality is mainly due to the variation of solar irradiance.

The LSTM model has been constructed and then used for forecasting. Data were split into training and testing datasets (90% and 10% respectively). Parameters of the constructed LSTM are presented in **Table 2**.

The evolution of the loss function and RMSE in the function of the number of iterations is displayed respectively in **Figures 10** and **11**. **Figures 10** and **11** indicate the convergence of the loss function and RMSE when the number of epochs is around 60.

Once the LSTM model has been constructed based on training data, it will be employed to forecast future values of power production. We attempt now to test

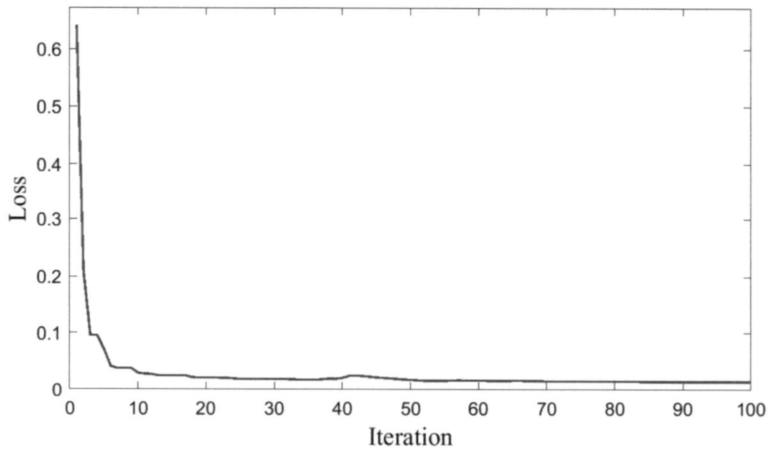

Figure 10.
Evolution of LSTM loss function during training stage.

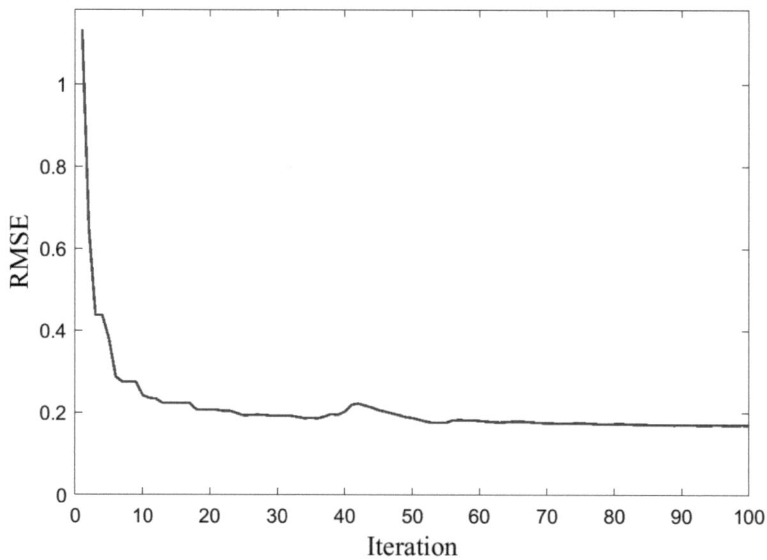

Figure 11.
Evolution RMSE of LSTM model during training stage.

the capability of the above LSTM model to forecast future values of the PV power generation. **Figure 12** shows the forecasting results of the PV power generation compared with the real data over a time horizon. **Figure 13** shows the scatter plot of the measured and forecasted power production via the LSTM model. It can be seen from **Figures 12** and **13** that the computed LSTM model has the ability of short-term forecasting of PV power generation. In addition, the forecasting result in **Figure 12** illustrates the efficiency of the LSTM model to forecast PV power production even under a cloudy day (i.e., the second day in **Figure 12**) where the power data is very dynamic.

In summary, the LSTM model showed good forecasting capacity with the coefficient of determination R^2 = 0.98 close to 1 and relatively small mean absolute

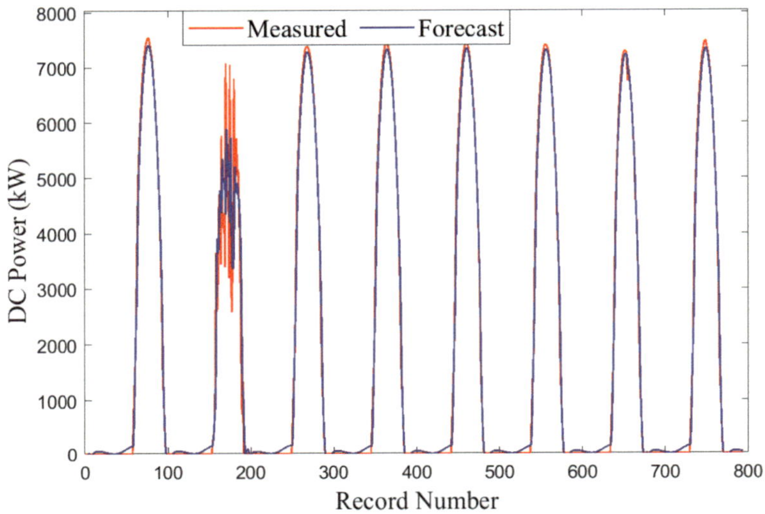

Figure 12.
Plot of collected solar power and forecasted one using LSTM model.

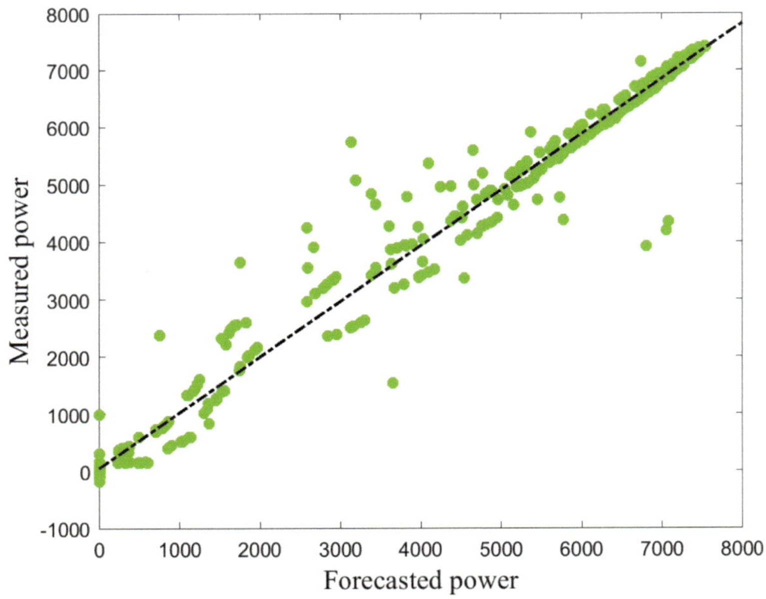

Figure 13.
Scatter graph of measured and LSTM forecast solar power output.

percentage error (MAPE), MAPE = 8.93. It should be pointed out that the forecasting accuracy in cloudy days could be improved by including meteorological variables, such as solar irradiance, ambient temperature, and wind velocity, as input variables.

4. Conclusion

The major challenge in solar energy generation is the volatility intermittent of photovoltaic system power generation due mainly to weather conditions. Thus, accurate forecasting of photovoltaic power generation is becoming indispensable for reducing the effect of uncertainty and energy costs and enable suitable integration of photovoltaic systems in a smart grid. This chapter employed a Long Short-Term Memory (LSTM) model to accurately forecast short-term photovoltaic solar power. This approach exploits the desirable properties of LSTM, which is a powerful tool for modeling dependency in data. The forecasting quality of this approach has been verified using data from January 2018 to December 2018 collected from a 9 MWp grid-connected plant. Promising results have been achieved by the proposed LSTM-based approach to short-term forecasting of photovoltaic solar power production. As future work, to further enhance the forecasting quality we plan to implement and test the performance of other RNN models like Gated recurrent unit (GRU) model and to incorporate other information such as meteorological data. Also, as most data from real plants are multiscale in nature and noisy, we plan in future work to merge the desirable LSTM model with the wavelet-based multiscale presentation [29]. This permits to get a multiscale LSTM model able to capture feature in both time and frequency and possess good ability to handle noisy data.

Acknowledgements

This publication is based upon work supported by King Abdullah University of Science and Technology (KAUST), Office of Sponsored Research (OSR) under Award No: OSR-2019-CRG7-3800.

Author details

Fouzi Harrou[1]*, Farid Kadri[2]* and Ying Sun[1]

1 Electrical and Mathematical Sciences and Engineering (CEMSE) Division, King Abdullah University of Science and Technology (KAUST) Computer, Saudi Arabia

2 Aeroline Customer Services, Sopra Steria Group, France

*Address all correspondence to: fouzi.harrou@kaust.edu.sa
and farid.kadri@soprasteria.com

References

[1] Sobri S, Koohi-Kamali S, Rahim NA. Solar photovoltaic generation forecasting methods: A review. Energy Conversion and Management. 2018;**156**: 459-497

[2] Harrou F, Taghezouit B, Sun Y. Improved kNN-based monitoring schemes for detecting faults in PV systems. IEEE Journal of Photovoltaics. 2019;**9**(3):811-821

[3] Antonanzas J, Osorio N, Escobar R, Urraca R, Martinez-de Pison F, Antonanzas-Torres F. Review of photovoltaic power forecasting. Solar Energy. 2016;**136**:78-111

[4] Eseye AT, Zhang J, Zheng D. Short-term photovoltaic solar power forecasting using a hybrid wavelet-PSO-SVM model based on SCADA and meteorological information. Renewable Energy. 2018;**118**:357-367

[5] Wang H, Yi H, Peng J, Wang G, Liu Y, Jiang H, et al. Deterministic and probabilistic forecasting of photovoltaic power based on deep convolutional neural network. Energy Conversion and Management. 2017;**153**:409-422

[6] Wang K, Qi X, Liu H. A comparison of day-ahead photovoltaic power forecasting models based on deep learning neural network. Applied Energy. 2019;**251**:113315

[7] Harrou F, Dairi A, Taghezouit B, Sun Y. An unsupervised monitoring procedure for detecting anomalies in photovoltaic systems using a one-class support vector machine. Solar Energy. 2019;**179**:48-58

[8] Harrou F, Taghezouit B, Sun Y. Robust and flexible strategy for fault detection in grid-connected photovoltaic systems. Energy Conversion and Management. 2019;**180**: 1153-1166

[9] Behera MK, Majumder I, Nayak N. Solar photovoltaic power forecasting using optimized modified extreme learning machine technique. Engineering Science and Technology, an International Journal. 2018;**21**(3): 428-438

[10] Qing X, Niu Y. Hourly day-ahead solar irradiance prediction using weather forecasts by LSTM. Energy. 2018;**148**:461-468

[11] Srivastava S, Lessmann S. A comparative study of LSTM neural networks in forecasting day-ahead global horizontal irradiance with satellite data. Solar Energy. 2018;**162**: 232-247

[12] Kushwaha V, Pindoriya NM. A SARIMA-RVFL hybrid model assisted by wavelet decomposition for very short-term solar PV power generation forecast. Renewable Energy. 2019;**140**: 124-139

[13] Persson C, Bacher P, Shiga T, Madsen H. Multi-site solar power forecasting using gradient boosted regression trees. Solar Energy. 2017;**150**: 423-436

[14] Lin K-P, Pai P-F. Solar power output forecasting using evolutionary seasonal decomposition least-square support vector regression. Journal of Cleaner Production. 2016; **134**:456-462

[15] Raza MQ, Mithulananthan N, Summerfield A. Solar output power forecast using an ensemble framework with neural predictors and Bayesian adaptive combination. Solar Energy. 2018;**166**:226-241

[16] Chen Z, Chen Y, Wu L, Cheng S, Lin P, You L. Accurate modeling of photovoltaic modules using a 1-d deep residual network based on iv

characteristics. Energy Conversion and Management. 2019;**186**:168-187

[17] Harrou F, Dairi A, Sun Y, Kadri F. Detecting abnormal ozone measurements with a deep learning-based strategy. IEEE Sensors Journal. 2018;**18**(17):7222-7232

[18] Harrou F, Dairi A, Sun Y, Senouci M. Statistical monitoring of a wastewater treatment plant: A case study. Journal of Environmental Management. 2018;**223**:807-814

[19] Dairi A, Harrou F, Sun Y, Senouci M. Obstacle detection for intelligent transportation systems using deep stacked autoencoder and k-nearest neighbor scheme. IEEE Sensors Journal. 2018;**18**(12):5122-5132

[20] Ugurlu U, Oksuz I, Tas O. Electricity price forecasting using recurrent neural networks. Energies. 2018;**11**(5):1255

[21] Fu R, Zhang Z, Li L. Using LSTM and GRU neural network methods for traffic flow prediction. In: 2016 31st Youth Academic Annual Conference of Chinese Association of Automation (YAC). Wuhan, China: IEEE; 2016. pp. 324-328

[22] Hochreiter S, Schmidhuber J. Long short-term memory. Neural Computation. 1997;**9**(8):1735-1780

[23] Dorffner G. Neural networks for time series processing. In: Neural Network World. Citeseer^{x}; 1996

[24] Oksuz I, Cruz G, Clough J, Bustin A, Fuin N, Botnar RM, et al. Magnetic resonance fingerprinting using recurrent neural networks. In: 2019 IEEE 16th International Symposium on Biomedical Imaging (ISBI 2019). Venice, Italy: IEEE; 2019. pp. 1537-1540

[25] Mukkamala MC, Hein M. Variants of RMSProp and Adagrad with logarithmic regret bounds. In: Proceedings of the 34th International Conference on Machine Learning-Volume 70. Sydney, Australia: JMLR.org; 2017. pp. 2545-2553

[26] Srivastava N, Hinton G, Krizhevsky A, Sutskever I, Salakhutdinov R. Dropout: A simple way to prevent neural networks from overfitting. The Journal of Machine Learning Research. 2014;**15**(1): 1929-1958

[27] Watt N, du Plessis MC. Dropout for recurrent neural networks. In: Oneto L, Navarin N, Sperduti A, Anguita D, editors. Recent Advances in Big Data and Deep Learning. Genova, Italy: Springer International Publishing; 2020. pp. 38-47

[28] Gal Y, Ghahramani Z. A theoretically grounded application of dropout in recurrent neural networks. In: Advances in Neural Information Processing Systems. Barcelona, Spain; 2016. pp. 1019-1027

[29] Harrou F, Sun Y, Madakyaru M. An improved wavelet-based multivariable fault detection scheme. In: Uncertainty Quantification and Model Calibration. IntechOpen; 2017. p. 175

Chapter 2

Optimal Design of Energy System Based on the Forecasting Data with Particle Swarm Optimization

Yamin Yan, Haoran Zhang, Jianqin Zheng and Yongtu Liang

Abstract

Renewable energy source has developed rapidly and attracted considerable attention. The integration of renewable energy into the energy supply chain requires precise forecast of the output of energy supply chain, thereby reducing energy resource waste and greenhouse gas emissions. In this study, a coupled model system is developed to forecast energy supply chain for the design optimization of distributed energy system, which can be divided into two parts. In the first part, long short-term memory (LSTM) and particle swarm optimization algorithm (PSO) contribute to energy supply chain forecast considering time series, and particle swarm optimization is used to optimize the parameters of the long short-term memory model to improve the forecast accuracy. Results show that the mean absolute error and root mean squared error are 8.7 and 16.3 for the PSO-LSTM model, respectively. In the second part, the forecast results are used as input of the distributed energy system to further optimize the design and operation schemes, so as to achieve the coupling optimization of forecast and design. Finally, a case study is carried out to verify the effectiveness of the proposed method.

Keywords: coupled model system, forecast, design optimization, renewable energy system

1. Introduction

With the development of social economy and the acceleration of urbanization and industrialization, the global energy consumption and greenhouse gas (GHG) emissions are expected to increase in the next few decades [1, 2]. At present, more than 80% of the world's primary energy comes from fossil fuels, and its carbon emissions are the biggest cause of global greenhouse effect. Therefore, the development of low-carbon economy renewable energy technologies is an emerging trend and goal of the world. In the past decade, renewable energy power generation has grown at an annual rate of 16%, and China accounted for 45% of global growth [3]. Wind power, one of the most potential renewable energy sources, has entered the rapid development stage. In 2018, the wind power has a largest share (around 50%) in the renewable energy power generation and reached 142 TW h [4], and more than 10% of electricity consumption or 5 GW of electricity will be generated by solar energy by 2020 [5]. As reported in "Clean Energy Trends" published in March 2011, the global production of biofuels, wind power, and solar energy is expected to reach $112.8 billion, $122.9 billion, and $113.6 billion by 2020, with growth rates of 116, 103, and 60%, respectively. Whether from the point of view of energy saving or environmental protection, renewable energy is certain to play a vital role in power generation [6, 7]. However, renewable energy has the characteristics of strong uncertainty and volatility, which not only increase the operation cost of power system but also reduce the cost-effectiveness of renewable energy resources [8]. Therefore, it is necessary to establish an accurate and stable prediction model for the renewable energy power to ensure the energy supply chain, so as to lay the foundation for optimizing the design and operation strategies of energy system [9, 10].

Many scholars have conducted the research on forecasting and optimizing the renewable energy supply system. Regarding the energy supply system forecast, many studies have developed effective forecasting methods for specific energy sources, which can be divided into three types of models, namely, time series models, data-driven models, and wavelet-based multiscale models [11]. Generally, the time series models and data-driven models are widely accepted. For instance, Lu et al. [12] compared the accuracy of four artificial intelligence methods in forecasting Taiwan's renewable energy sources based on historical data from 2000 to 2015, and results showed that only grey forecasting model coupled with heuristic fuzzy time series method is suitable for small dataset forecast. Long et al. [13] used four famous data-driven method to forecast daily solar power after dimensionality reduction of data, and they inferred that different algorithms can outperform others in different considered scenarios. There are few literatures on the application of the wavelet-based multiscale models. Reikard [14] proposed a Kalman filter and time-varying regression for the time series prediction of wave energy. Aasim et al. [15] proposed a new repeated wavelet transform based ARIMA model for very short-term wind speed forecasting. Based on data-driven, Li et al. [16] developed a new deep machine learning algorithm to predict short-term wave energy. In summary, different forecast models may vary greatly in different scenarios.

However, most of the previous work has been devoted to the forecast of renewable energy in a certain region considering the limited range of measuring instruments, and few studies pay attention to the power generation in multiple regions. In fact, it is impossible to install the measuring instruments in each region from an economic point of view, thus bringing a great challenge for forecasting renewable energy in multiple regions. In addition, the forecast of renewable energy is not only related to the energy intensity but also depends on climate conditions, which have a strong time series relationship [17]. Meanwhile, the short stochastic characteristics and the dependence of the observation time series of renewable energy must be taken into account. Therefore, this paper is the first work to develop a forecast model based on the data-driven method by collecting the measurement equipment data and climate conditions in different regions and at different times. Also, the LSTM model is adopted to capture the dependence of time series forecasting for renewable energy output, and PSO is used to optimize the parameters of the LSTM model to improve the forecast accuracy.

Regarding the optimization of renewable energy supply system, many scholars are devoted to integrating various renewable energy sources and have made a range of achievements [18–20]. Acevedo-Arenas et al. [21] described the methodology of hybridizing photovoltaic, wind, and forest biomass energy sources and developed a model to optimize the design and operation schemes of hybrid renewable energy system in consideration of economic and environmental impacts. Tajeddin and Roohi [22] put forward the use of biomass energy to improve the responsibility and efficiency of wind farms. Sakaguchi and Tabata [23] forecasted the power generation potential and proportion of biomass, wind power, and PV in Awaji Island Japan, and results showed that it is possible to achieve self-sufficiency by the sole use of renewable energy. Moreover, the renewable energy plants are usually equipped with the energy storage, which could improve the controllability of the output [24]. Based on the above research, the hybridization of renewable energy sources is able to reduce the uncertainty and volatility, thereby promoting the rational use of renewable energy and increasing environmental benefits.

In this paper, we first focus on the accurate forecast of renewable energy supply chain, which is divided into three steps. The first step is to develop a forecast model based on the data-driven method by collecting the measurement equipment data and climate conditions in different regions and at different times. In the second step, the LSTM model is adopted to capture the dependence of time series forecasting for renewable energy output. In the third step, PSO is used to optimize the parameters of the LSTM model to improve the forecast accuracy. After that, the forecast results are used as input of the distributed energy system to further optimize the design and operation schemes, thereby achieving the coupling optimization of forecast and design. Overall, there are four contributions in this paper:

1. LSTM model is established to forecast the output of renewable energy in multiple regions, considering the correlation of time series and the change of climate conditions.

2. PSO is adopted to optimize the parameters of LSTM model, thereby improving the forecast accuracy.

3. Based on the forecast results of LSTM model, a MILP model is established to optimize the design and operation schemes of renewable energy supply system.

4. Real multi-region renewable energy plants are used as an example to verify the effectiveness and practicality of the proposed coupled model system.

The rest of this paper is organized as follows: Section 2 describes the basic knowledge of LSTM model and PSO algorithm, and the optimization model of distributed renewable energy supply system is explained. Subsequently, the optimal results solved by the coupled model system and corresponding discussions are presented in Section 3. Finally, Section 4 concludes this paper.

2. Problem formulation and solution methodology

2.1 LSTM model coupled with PSO algorithm

Figure 1 shows the framework of the coupled model system. Firstly, long short-term memory (LSTM) and particle swarm optimization algorithm (PSO) contribute

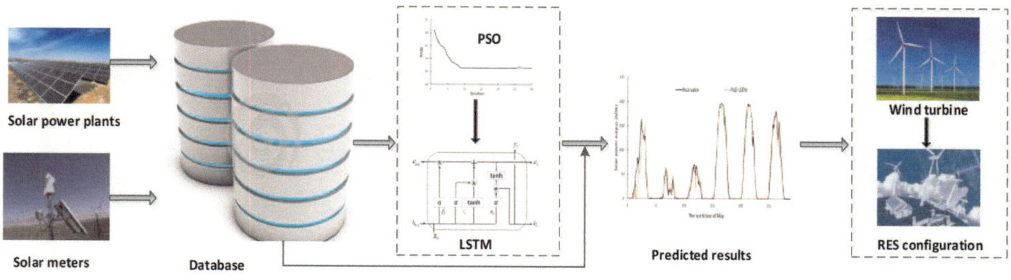

Figure 1.
The framework of the coupled model system.

to energy supply chain forecast considering time series, and particle swarm optimization is used to optimize the parameters of the long short-term memory model to improve the forecast accuracy. Then, the forecast results of solar power plants and the known wind power are used as input of the distributed energy system to further optimize the design and operation schemes, so as to achieve the coupling optimization of forecast and design.

2.2 LSTM model coupled with PSO algorithm

2.2.1 LSTM model

As a kind of recurrent neural network, LSTM model combines short-term with long-term memory through subtle gate control and overcomes the shortcoming of gradient vanishing. Currently, LSTM model has great applications in time series, such as machine translation and speech recognition. In this section, the basic LSTM cell is introduced firstly.

The basic LSTM cell is shown in **Figure 2**. At time t, the cell state at t−1, the hidden state at t−1, and the new information are denoted as $c_{t\text{-}1}$, $h_{t\text{-}1}$, and x_t, respectively, which are the input of the cell. Forget gate, input gate, and output gate are used to select information and depended on x_t and $h_{t\text{-}1}$. The output value of each gate is limited to between 0 and 1 by the activation function, which is the sigmoid function defined in Eq. (1).

$$\sigma(x) = \frac{1}{1 + e^{-x}} \tag{1}$$

The LSTM cell included four steps as follows. Firstly, the cell decides what stored information is going to be thrown away from the previous cell state $c_{t\text{-}1}$. Based on x_t and $h_{t\text{-}1}$, forget gate f_t outputs a value between 0 and 1 for $c_{t\text{-}1}$. And 1 means that the stored information is completely retained, and 0 means that the information is completely eliminated. f_t is calculated by Eq. (2).

$$f_t = \sigma\left(W_{xf} \cdot x_t + W_{hf} \cdot h_{t\text{-}1} + b_f\right) \tag{2}$$

Secondly, the cell decides what new information is going to be stored. At first, tanh function defined in Eq. (3) is to ensure the normalization of the new information and converts x_t and $h_{t\text{-}1}$ to the new format between −1 and 1 denoted as in Eq. (4). Then, the input gate i_t defined in Eq. (5) outputs a value between 0 and 1 to

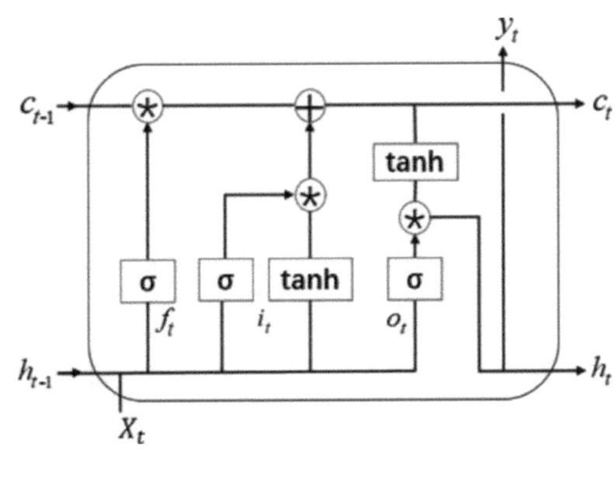

Figure 2.
The basic LSTM cell.

select the new information. After that, these two parts are going to be combined to update the cell state in the next step.

$$\tanh(x) = \frac{e^x - e^{-x}}{e^x + e^{-x}} \tag{3}$$

$$\overline{c_t} = \tanh(W_{xc} \cdot x_t + W_{hc} \cdot h_{t\text{-}1} + b_c) \tag{4}$$

$$i_t = \sigma(W_{xi} \cdot x_t + W_{hi} \cdot h_{t\text{-}1} + b_i) \tag{5}$$

Thirdly, it is going to update the cell state based on the previous steps. The old cell state is multiplied by f_t, forgetting the information decided to be forgotten in step 1. The process information $\overline{c_t}$ decided by tanh function is multiplied by i_t. Then these two parts are added to determine the new cell state c_t in Eq. (6).

$$c_t = f_t * c_{t\text{-}1} + i_t * \overline{c_t} \tag{6}$$

Finally, the cell decides what is going to be the output. Similar to other gates, output gate o_t outputs a value to decide what parts of the cell state are going to be output in Eq. (7). Based on c_t, the hidden state h_t defined in Eq. (8) is multiplied by o_t to decide the important information to be stored. At last, the output of the LSTM cell is shown in Eq. (9).

$$o_t = \sigma(W_{xo} \cdot x_t + W_{ho} \cdot h_{t\text{-}1} + b_o) \tag{7}$$

$$h_t = o_t * \tanh(c_t) \tag{8}$$

$$y_t = \sigma(W_{hy} \cdot h_t + b_y) \tag{9}$$

In this work, X is the measurement equipment data and climate conditions, and y is the output of the renewable energy. W_{xf}, W_{xi}, W_{xo}, and W_{xc} are the input weight matrices; W_{hf}, W_{hi}, W_{ho}, and $W_{hc}$are the recurrent weight matrices; and W_{hy} is the hidden output weight matrix. Vectors b_f, b_i, b_o, b_c, and b_y are the corresponding bias vectors.

2.2.2 PSO algorithm

Proposed in 1995, PSO has a great use in the field of optimization, especially coupling with machine learning, such as ANN [25]. In this study, PSO is used to select the optimal parameters for the LSTM model. The process of the algorithm is described as follows.

Firstly, initializing the particle swarm, including the size of population and the dimension of each one. In this work, the number of units, learning rate, and time step are optimized, so the dimension of each individual is 3.

Secondly, evaluating the fitness of each individual. In the process of training the LSTM model, the training loss is the fitness of each particle.

Thirdly, finding the two "extreme value." In each iteration, by calculating the training loss, the individual optimal value and optimal population value are found, which are used for updating the particles.

Fourthly, updating the velocity and position of each particle. Based on the two "extreme value," the particles are updated to form the next generation according to Eqs. (10) and (11):

$$v_i(t+1) = w \times v_i + c_1 \times r_1 \times (p_i(t) - x_i(t)) + c_2 \times r_2 \times \left(p_g(t) - x_i(t)\right) \tag{10}$$

$$x_i(t+1) = x_i(t) + v_i(t+1) \tag{11}$$

where v_i and x_i are the velocity and position of particle i, respectively. p_i is the best position for a particle i and p_g denotes the best position in the group at the tth iteration.

Fifthly, reaching convergence condition. With the help of iterations, the best particle will be found within the maximum iteration. So the suitable parameters of the LSTM model are determined.

The methodology of LSTM model coupled with PSO algorithm is composed of three parts. Firstly, the data of measurement equipment data and climate conditions are collected into the database with preprocessing. Secondly, the data is divided into two groups, and small part of the data is for the PSO algorithm to train the LSTM parameters. It contains the process of PSO algorithm mentioned in 3.1.2 section. Finally, after finding the optimal LSTM model, train the prediction model with the rest of the data. So far, a prediction model with high accuracy for the output of the renewable energy is established.

2.3 The optimization model of energy supply system

2.3.1 Objective function

The objective of this model is to minimize the total annual cost of the energy supply system, C_{TOT}, defined as the sum of the total fixed cost, capital cost, and energy consumption cost. It should be noted that the input energy of the energy supply system is renewable energy sources; thus, the energy consumption cost is equal to zero:

$$C_{TOT} = C_{Cac} + C_F \tag{12}$$

where C_{Cac} and C_F are the total capital cost and fixed cost of the energy supply system, including the PV panels, wind turbines, and batteries [CNY].

The annual capital cost of technology i is obtained by the capital recovery factor CRF_i, which is defined as Eq. (15):

$$C_{Cac} = Cap_i C_{UCi} CRF_i \tag{13}$$

$$C_F = B_i C_{FCi} CRF_i \tag{14}$$

$$CRF_i = \frac{r(1+r)^{N_i}}{(1+r)^{N_i} - 1} \tag{15}$$

where Cap_i is the rated power of technology i [kW, kWh]; B_i is the binary variable, which is equal to 1 if the technology i is selected for installation [−]; C_{UCi} represents the unit cost of technology i [CNY/kW for energy generation technology and CNY/kWh for energy storage technology]; C_{FCi} is the fixed cost of technology i [CNY]; N_i represents the lifetime of technology i [year]; and r defines the discount rate [−].

2.3.2 Constraints

In this section, three types of constraints are considered: energy balances, design, and operation constraints. Regarding the energy balance constraints, the electricity demand is satisfied by the PV panels and wind turbines. When the electric load changes, the batteries are used to effectively adjust the electricity balance and keep the voltage and frequency constant, and Eqs. (17) and (18) are required to describe the operation state of batteries. Notably, the batteries are only able to deal with the daily fluctuations, and Eq. (16) can be applied to any hour except for the first hour of the day, while Eq. (18) can only be applied to the first hour of the day:

$$Q_{t,PV} + Q_{t,WT} + S_t^{dis} - S_t^{ch} = D_t \tag{16}$$

$$S_t = S_{t-1}n_{loss} + S_t^{ch}n_{ch} - S_t^{dis}/n_{dis} \tag{17}$$

$$S_t = S_{t+23}n_{loss} + S_t^{ch}n_{ch} - S_t^{dis}/n_{dis} \tag{18}$$

where D_t is the electric load at time-window t [kWh]; $Q_{t,PV}$ and $Q_{t,WT}$ indicate the energy output from the PV panels and wind turbines at time-window t [kWh]; S_t^{ch} and S_t^{dis} are the charging and discharging rate of the battery at time-window t [kWh]; S_t represents the electricity stored in the battery at time-window t [kWh]; and n_{loss}, n_{ch}, and n_{dis} represent the battery's loss, charging, and discharging efficiency, respectively, [−].

As for the design and operation constraints, the operating power of technology i should be less than the rated power, as defined in Eqs. (19) and (20). In addition, Eq. (21) requires that the installation capacity must be within the maximum limit to ensure safe and stable operation of technology i:

$$Q_{t,i} \le Cap_i \tag{19}$$

$$S_t \le Cap_{battery} \tag{20}$$

$$0 \le Cap_i \le B_i Max_i \tag{21}$$

where Max_i is the upper limit of the capacity of technology i [kW, kWh].

2.3.3 Model solving

Based on the computer with 8 GB of RAM memory and solution environment with 1.6 GHz of CPU, the model was programmed with the software MATLAB R2015 and solved by the business solver GUROBI 8.1.0 to obtain the global optimum.

3. Case study

3.1 The forecast results of LSTM model

Firstly, this part is to predict the solar energy of the solar power plants. In area A, there are eight solar plants located in different regions. Historical data from January to May 2017 is used for experiment with a time interval of 30 minutes. The measurement equipment data including measured temperature and global solar radiation is from the power system, and the weather conditions are from the weather bureau. In this study, in order to evaluate the quality of the LSTM model, the evaluation indexes are root mean squared error (RMSE) and mean absolute error (MAE). The formulas of these two are defined as follows:

$$MAE = \frac{1}{n}\sum_{i=1}^{n} |p_i - r_i| \tag{22}$$

$$RMSE = \sqrt{\frac{1}{n}\sum_{i=1}^{n}(p_i - r_i)^2} \tag{23}$$

where n means the total number of the data, p_i is the predicted value, and r_i is the real value.

In this part, the data of January is used for training PSO-LSTM. The training sets, validation sets, and testing sets account for 80, 10, and 10%. And RMSE of the testing data is used as the fitness value of PSO. The parameters of the LSTM model are hidden units, learning rate, and time step. The corresponding range is 1–30, 0.0001–0.1, and 1–50 respectively, as shown in **Table 1**. After executing the PSO-LSTM, the suitable parameters are optimized and shown in **Table 1**. The convergence of RMSE is figured in **Figure 3**.

After determining the suitable LSTM model, the prediction model for solar energy is established. Data from February to mid-May is used to train this predictive model, and other data is used to test its accuracy. After the data is tested, the MAE and RMSE are 8.7 and 16.3, respectively. In order to illustrate the superiority of PSO-LSTM, the prediction results of basic LSTM and artificial neural network (ANN) are compared. And the predicted results and real value of the last 6 days of May are shown in **Figure 4**.

3.2 The optimal results of the renewable energy system

From the aforementioned analysis, the prediction model for solar energy is accurate. With the input of the measurement equipment data and weather conditions, the output of the solar energy can be predicted precisely. The hourly output forecast curve of the solar power plant and the known wind power energy and the electricity demand are shown in **Figure 5**. The technical and cost information of

	Neural units	**Learning rate**	**Time step**
Range	1–30	0.0001–0.1	1–50
Result	20	0.01	5

Table 1.
The parameters of the LSTM network.

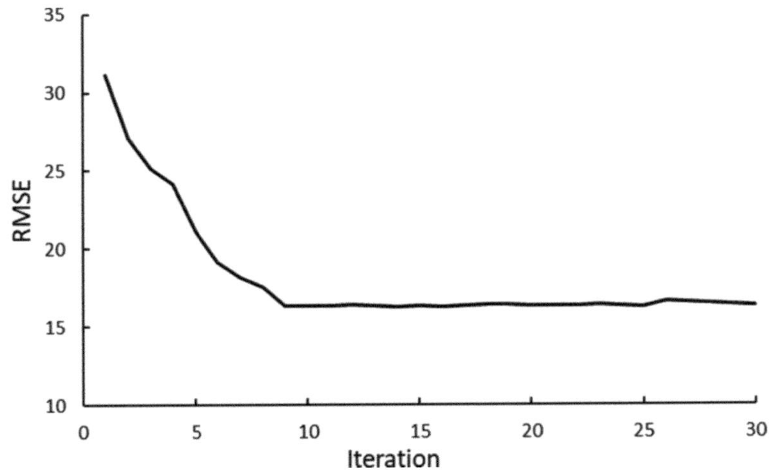

Figure 3.
RMSE of the PSO-LSTM.

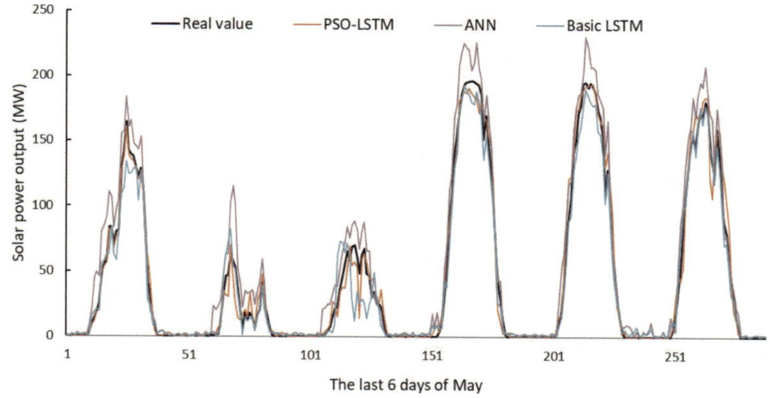

Figure 4.
The predicted results and the real value.

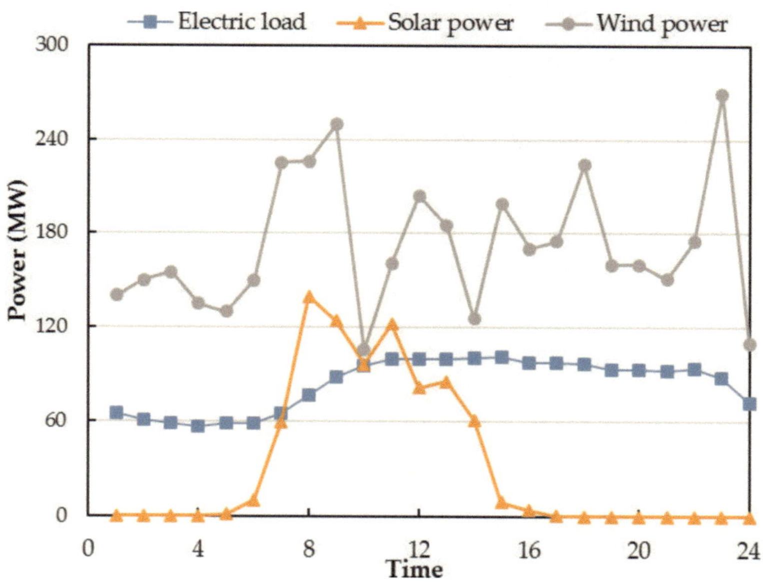

Figure 5.
Forecast of solar power plant and wind farm and electricity demand.

technologies are summarized in **Table 2** [26–28], and the loss, charging, and discharging efficiency of battery are set as 99.9, 90, and 90%, respectively [29]. The discount rate is assumed to be 8% to annualize the total investment cost.

By solving the MILP model, the optimal results of the renewable energy system can be obtained. The annual total cost is equal to 109.91 million. Specifically, the

Technology	Size range (MW, MWh)	Efficiency	Capital cost (CNY/kW, CNY/kWh)	Fixed cost (CNY)	Lifetime (a)
PV panels	0–500	17%	1500	34,500	20
Wind turbine	0–500	35%	3500	17,900	25
Battery	—	—	85	11,600	20

Table 2.
The technical and cost information of technologies.

total capacities of PV panels, wind turbines, and batteries are 140, 270, and 185 MWh. Obviously, the PV panels are selected due to the low investment cost, and the capacities reach the optimal to match with the limited solar power. The wind turbines are installed to work when the solar power is equal to 0. The batteries are installed to store the energy generated by the energy generation technologies to reduce the investment cost.

4. Conclusion

In this work, LSTM model is established to forecast the output of renewable energy in multiple regions, considering the correlation of time series and the change of climate conditions. Then, PSO algorithm is applied to optimize the parameters of the LSTM model. Results show that PSO-LSTM model is highly accurate with small error. Based on the forecast results, a MILP model is established to obtain the configuration of the renewable energy system with minimal total annual cost.

In future work, other intelligent algorithms, such as differential evolution, the simulated annealing algorithm, and ant colony optimization, can also be applied to select suitable parameters for long short-term memory. Besides, it is also important to study solar and wind energy access to power systems.

Acknowledgements

This work was partially supported by the National Natural Science Foundation of China (51874325) and the Grant-in-Aid for Early-Career Scientists (19 K15260) from the Japan Ministry of Education, Culture, Sports, Science and Technology. The authors are grateful to all study participants and, additionally, thanks to the solar power output data provided by the "PV in HOKKAIDO" contest hosted by Tokyo Electric Power Company Holdings (TEPCO) and Hokkaido Electric Power Company (HEPCO).

Appendices and nomenclature

Sets and indices

$i \in I$	all technologies: PV panels, wind turbine, and battery
$t \in T$	number of time windows

Continuous parameters

C_{Cac}	total capital cost [CNY]
C_F	total fixed cost [CNY]
C_{FCi}	fixed cost of technology i [CNY]
C_{UCi}	unit cost of technology i [CNY/kW, CNY/kWh]
CRF_i	capital recovery factor of technology i [-]
c_t	the cell state at time t
$\overline{c_t}$	the new candidate information at time t
D_t	electric load at time-window t [kWh]
f_t	the forget gate at time t
h_t	the hidden state at time t
i_t	the input gate at time t
Max_i	the upper limit of the capacity of technology i [kW, kWh]

n_{loss}	the loss efficiency of the battery [-]
n_{ch}	the charging efficiency of the battery [-]
n_{dis}	the discharging efficiency of the battery [-]
N_i	lifetime of technology i [year]
o_t	the output gate at time t
$Q_{t,i}$	energy output from the technology i, at time-window t [kWh]
x_t	the input information at time t

Positive variables

B_i	if the technology i is selected for installation, $B_i = 1$; otherwise, $B_i = 0$
Cap_i	rated power of technology i [kW, kWh]
S_t^{ch}	charging rate of the battery at time-window t [kWh]
S_t^{dis}	discharging rate of the battery at time-window t [kWh]
S_t	electricity stored in the battery at time-window t [kWh]

Author details

Yamin Yan[1], Haoran Zhang[2*], Jianqin Zheng[1] and Yongtu Liang[1]

1 Beijing Key Laboratory of Urban Oil and Gas Distribution Technology, China University of Petroleum-Beijing, Beijing, China

2 Center for Spatial Information Science, The University of Tokyo, Kashiwa, Chiba, Japan

*Address all correspondence to: zhang_ronan@csis.u-tokyo.ac.jp

References

[1] Ürge-Vorsatz D, Cabeza LF, Serrano S, Barreneche C, Petrichenko K. Heating and cooling energy trends and drivers in buildings. Renewable and Sustainable Energy Reviews. 2015;**41**: 85-98. DOI: 10.1016/j.rser.2014.08.039

[2] Kannan N, Vakeesan D. Solar energy for future world—A review. Renewable and Sustainable Energy Reviews. 2016; **62**:1092-1105. DOI: 10.1016/j.rser.2016.05.022

[3] BP. Statistical Review of World Energy. 2019. Available from: https://www.bp.com/content/dam/bp/business-sites/en/global/corporate/pdfs/energy-economics/statistical-review/bp-stats-review-2019-full-report.pdf

[4] Yang FF, Zhao XG. Policies and economic efficiency of China's distributed photovoltaic and energy storage industry. Energy. 2018;**154**: 221-230. DOI: 10.1016/j.energy.2018.04.135

[5] Tsai S-B, Xue Y, Zhang J, Chen Q, Liu Y, Zhou J, et al. Models for forecasting growth trends in renewable energy. Renewable and Sustainable Energy Reviews. 2017;**77**:1169-1178. DOI: 10.1016/j.rser.2016.06.001

[6] Treyer K, Bauer C. The environmental footprint of UAE's electricity sector: Combining life cycle assessment and scenario modeling. Renewable and Sustainable Energy Reviews. 2016;**55**:1234-1247. DOI: 10.1016/j.rser.2015.04.016

[7] Al-Karaghouli A, Kazmerski LL. Energy consumption and water production cost of conventional and renewable-energy-powered desalination processes. Renewable and Sustainable Energy Reviews. 2013;**24**:343-356. DOI: 10.1016/j.rser.2012.12.064

[8] Lin S, Li C, Xu F, Liu D, Liu J. Risk identification and analysis for new energy power system in China based on D numbers and decision-making trial and evaluation laboratory (DEMATEL). Journal of Cleaner Production. 2018; **180**:81-96. DOI: 10.1016/j.jclepro.2018.01.153

[9] Wang F, Zhang Z, Liu C, Yu Y, Pang S, Duić N, et al. Generative adversarial networks and convolutional neural networks based weather classification model for day ahead short-term photovoltaic power forecasting. Energy Conversion and Management. 2019;**181**:443-462. DOI: 10.1016/j.enconman.2018.11.074

[10] Li Q, Loy-Benitez J, Nam K, Hwangbo S, Rashidi J, Yoo C. Sustainable and reliable design of reverse osmosis desalination with hybrid renewable energy systems through supply chain forecasting using recurrent neural networks. Energy. 2019;**178**:277-292. DOI: 10.1016/j.energy.2019.04.114

[11] Qin W, Wang L, Lin A, Zhang M, Xia X, Hu B, et al. Comparison of deterministic and data-driven models for solar radiation estimation in China. Renewable and Sustainable Energy Reviews. 2018;**81**:579-594. DOI: 10.1016/j.rser.2017.08.037

[12] Lu S-L. Integrating heuristic time series with modified grey forecasting for renewable energy in Taiwan. Renewable Energy. 2019;**133**:1436-1444. DOI: 10.1016/j.renene.2018.08.092

[13] Long H, Zhang Z, Su Y. Analysis of daily solar power prediction with data-driven approaches. Applied Energy. 2014;**126**:29-37. DOI: 10.1016/j.apenergy.2014.03.084

[14] Reikard G. Forecasting ocean wave energy: Tests of time-series models. Ocean Engineering. 2009;**36**:348-356. DOI: 10.1016/j.oceaneng.2009.01.003

[15] Aasim SSN, Mohapatra A. Repeated wavelet transform based ARIMA model for very short-term wind speed forecasting. Renewable Energy. 2019; **136**:758-768. DOI: 10.1016/j.renene.2019.01.031

[16] Li L, Yuan Z, Gao Y. Maximization of energy absorption for a wave energy converter using the deep machine learning. Energy. 2018;**165**:340-349. DOI: 10.1016/j.energy.2018.09.093

[17] Bin Shams M, Haji S, Salman A, Abdali H, Alsaffar A. Time series analysis of Bahrain's first hybrid renewable energy system. Energy. 2016; **103**:1-15. DOI: 10.1016/j.energy.2016.02.136

[18] Akbari E, Hooshmand R-A, Gholipour M, Parastegari M. Stochastic programming-based optimal bidding of compressed air energy storage with wind and thermal generation units in energy and reserve markets. Energy. 2019;**171**:535-546. DOI: 10.1016/j.energy.2019.01.014

[19] Sarkar T, Bhattacharjee A, Samanta H, Bhattacharya K, Saha H. Optimal design and implementation of solar PV-wind-biogas-VRFB storage integrated smart hybrid microgrid for ensuring zero loss of power supply probability. Energy Conversion and Management. 2019;**191**:102-118. DOI: 10.1016/j.enconman.2019.04.025

[20] Vasilj J, Sarajcev P, Jakus D. Estimating future balancing power requirements in wind–PV power system. Renewable Energy. 2016;**99**: 369-378. DOI: 10.1016/j.renene.2016.06.063

[21] Acevedo-Arenas CY, Correcher A, Sánchez-Díaz C, Ariza E, Alfonso-Solar D, Vargas-Salgado C, et al. MPC for optimal dispatch of an AC-linked hybrid PV/wind/biomass/H2 system incorporating demand response. Energy Conversion and Management. 2019;**186**: 241-257. DOI: 10.1016/j.enconman.2019.02.044

[22] Tajeddin A, Roohi E. Designing a reliable wind farm through hybridization with biomass energy. Applied Thermal Engineering. 2019;**154**: 171-179. DOI: 10.1016/j.applthermaleng.2019.03.088

[23] Sakaguchi T, Tabata T. 100% electric power potential of PV, wind power, and biomass energy in Awaji island Japan. Renewable and Sustainable Energy Reviews. 2015;**51**:1156-1165. DOI: 10.1016/j.rser.2015.06.056

[24] Appino RR, González Ordiano JÁ, Mikut R, Faulwasser T, Hagenmeyer V. On the use of probabilistic forecasts in scheduling of renewable energy sources coupled to storages. Applied Energy. 2018;**210**:1207-1218. DOI: 10.1016/j.apenergy.2017.08.133

[25] Zheng J, Zhang H, Yin L, Liang Y, Wang B, Li Z, et al. A voyage with minimal fuel consumption for cruise ships. Journal of Cleaner Production. 2019;**215**:144-153. DOI: 10.1016/j.jclepro.2019.01.032

[26] Yan Y, Zhang H, Long Y, Wang Y, Liang Y, Song X, et al. Multi-objective design optimization of combined cooling, heating and power system for cruise ship application. Journal of Cleaner Production. 2019;**233**:264-279. DOI: 10.1016/j.jclepro.2019.06.047

[27] Xing X, Yan Y, Zhang H, Long Y, Wang Y, Liang Y. Optimal design of distributed energy systems for industrial parks under gas shortage based on augmented ε-constraint method. Journal of Cleaner Production. 2019;**218**:782-795. DOI: 10.1016/j.jclepro.2019.02.052

[28] Di Somma M, Yan B, Bianco N, Graditi G, Luh PB, Mongibello L, et al. Multi-objective design optimization of distributed energy systems through cost and exergy assessments. Applied

Energy. 2017;**204**:1299-1316. DOI: 10.1016/j.apenergy.2017.03.105

[29] Mavromatidis G, Orehounig K, Carmeliet J. Uncertainty and global sensitivity analysis for the optimal design of distributed energy systems. Applied Energy. 2018;**214**:219-238. DOI: 10.1016/j.apenergy.2018.01.062

Chapter 3

Neutral Network Adaptive Filter with Application to Ocean Current Energy Estimation

Hong Son Hoang and Remy Baraille

Abstract

This chapter proposes a new approach for the design of an adaptive filter (AF), which is based on an artificial neural network (NN) structure for estimating the system state. The NNs are now widely used as a technology offering a way to solve complex and nonlinear problems such as time-series forecasting, process control, parameter state estimation, and fault diagnosis. The proposed NN-based adaptive filtering (NNAF) is designed by considering the filtering algorithm as an input–output system and two-stage optimization procedure. The first concerns a learning process where the weights of the NNAF are estimated to minimize the error between the filtered state and the state samples generated by a numerical model. The adaptation is carried out next to minimize the mean prediction error (MPE) of the system outputs (error between the observations and the system output forecast) subject to the coefficients associated with the estimated NN weights. Simulation results for different numerical models, especially for state estimation of the chaotic Lorenz system as well as for the ocean current at different deep layers which is important for renewable energy device placements, are presented to show the efficiency of the NNAF.

Keywords: neural network, Kalman filter, adaptive filter, machine learning, minimum prediction error approach, ocean current energy

1. Introduction

In this chapter, we develop an NNAF for solving filtering problems, especially for dynamic systems of high dimension (an order of 10^7–10^8).

Optimal filtering originated from the astronomical studies at the end of the eighteenth century. It is worth of mentioning the early works on the filtering [1, 2], where the theory of linear minimum-variance filters is developed independently. In Ref. [1], the filter is developed to estimate the desired value of a noisy signal, whereas the filter in Ref. [2] minimizes the mean squared error between the estimated random process and the desired process. These two works constitute a foundation for the celebrated Kalman filter (KF) [3], which processes the measurements streaming in continuously and updates the estimate of the signal recursively upon the arrival of each measurement.

Under the condition that the system state and observation processes satisfy linear equations driven by white Gaussian noises, the KF produces the unbiased

IntechOpen

estimate with minimum error variance. Since in most applications, the dynamic and observation processes are nonlinear, and a linearization technique is applied, which results in an extended Kalman filter (EKF). However, the EKF is sometimes far from optimal and even divergent.

The KF has received a great interest for many industrial and engineering fields since the 1970s, such as the system control, signal processing, robotics, seismology, communication, economics, and finance.

Direct application of the KF to the data assimilation (DA) in geophysical systems such as meteorological or oceanic systems is very difficult, even impossible due to high computational effort and memory storage required to deal with very high system dimensions, not to say on nonlinearities and uncertainties in statistics of the model and observation errors.

This situation forces researchers to search new, simple but effective algorithms to overcome these difficulties. We find here the approaches more appropriate for solving DA problems in high dimensional systems (HdSs). For example, the 4D-Var algorithm [4] adjusts a forecast, in space and time, to bring it into closer agreement with observations. Based on the KF version, the EnKF [5] uses the covariance estimated from samples instead of solving the Algebraic Riccati matrix equation. The EnKF is related to the particle filter (a particle is the same thing as ensemble member). Particle filtering [6] uses a set of particles (also called samples) to represent the posterior distribution of some stochastic process given noisy and/or partial observations. The state-space model can be nonlinear, and the initial state and noise distributions can be non-Gaussian. These approaches are proposed with strategies to better control the uncertainty and to make it less non-Gaussian: targeting of observations, specific treatment of highly nonlinear degrees of freedom, localization, and model reduction. A review of the application and usefulness of 4D-Var, ensemble KF, and particle filters in geosciences is given in Ref. [7]. Among the efficient approaches for solving DA in high dimensional setting, we note the adaptive filtering (AF, see Section 2.2). The idea of the AF will be exploited in this chapter to improve the neural network filtering (NNF) algorithms.

To make more clear on why the AF is capable of dealing with the difficulties to assimilate data into the HdSs, it is useful to list the main differences between the KF and the AF as follows:

1. The KF is aimed at seeking an optimal estimate in the space of measurable Borel functions of the system state. As for the AF, it seeks the estimate, optimal in the class of stable filters of a given structure.

2. Optimality: the KF optimizes the mean squared error between the estimate and true system state given observations up to and included the recent moment k. The AF minimizes the mean squared error between the observation and the forecast of system output.

3. The parameters to be estimated in the KF are all components of the system state, which are time varying. The tuning parameters in the AF are the parameters in the filter gain, which are assumed to be time invariant. It is mentioned that the tuning parameters in the AF have no physical sense.

4. Nonlinear matrix equations are to be solved in the KF.

In this chapter, a new NNAF is proposed and tested for solving DA problems with moderate and high dimensional systems. In particular, the problem of estimating the ocean underwater current by the NNAF will be presented in detail due to its importance in producing energy from the ocean. It is mentioned that there are some basic ways to tap the ocean for its energy such as ocean wave, tidal range, tidal current, underwater current, and temperature differences in water. All these variables can be estimated by the NNAF. Ocean currents are suitable locations for deploying energy extraction devices such as turbines.

At the present, the use of ocean energy remains mainly in the prototype stage. Last few years, the prototypes are no longer confined only to laboratories, and some companies have set up commercial scale like RITE Project. In 2006, Verdant Power installed its first grid-connected Free Flow turbine (see Ref. [8]).

The modern DA techniques represent a computational challenge, even with the use of parallel computing with thousands of processors. Today, many operational weather prediction centers use high resolution models, with the number of observations (provided by a new satellite generation) growing exponentially. There is an increasing need to address new DA algorithms for HdSs and to improve their performance. This makes the problem of developing new DA algorithms like NNAF a great computational challenge.

In this chapter, we propose a new version of an artificial NN-based AF as a new DA technique, with the objective to construct a high performance forecasting system. Recently, the NN has become a popular and helpful model for classification, clustering, pattern recognition, and prediction in many disciplines. For a survey of NN applications in the real-world scenario, see Ref. [9]. In Ref. [10], the focus is on explaining the concept and evolution of machine learning, and some of the popular machine learning algorithms are described and compared. The work of Ref. [11] describes the important problems that can arise when applying NN models to meteorology and oceanography, especially in large-scale, low-frequency studies, and the attempts to resolve these problems. For data assimilation in chaotic dynamics using NNs, see Ref. [12].

In this chapter, the NNF based on AF developed in Ref. [13] will be constructed and implemented to produce the estimation results, which will serve as the reference for performance comparison with other NNFs. The need in developing the AF is to construct a filter, which is suitable in a new environment. The AF is a powerful device for signal processing, control applications, identification, inverse modeling, and prediction in an uncertain environment. The AF is formally defined as a *self-designing device with time-varying parameters that are adjusted recursively* in order to minimize an error function—a distance measurement between the reference or desired signal and the output of the adaptive filter. For a review of the AF, see Ref. [14].

The idea on the AF for data assimilation in HdS has been first presented in Ref. [13]. The originality of this AF approach lies in considering the AF as a filter, which minimizes the MPE for the system output, with tuning variables chosen as some pertinent parameters of the filter gain. This idea allows to optimize the filter performance iteratively, in real time, but not on the basis of a batch of observations as done in the 4D-Var approach. Moreover, the *dimensionless* tuning parameters are chosen in a way to ensure a stability of the filter. This makes the filter always stable during an optimization process and easily controls the values of tuning parameters. For a detail presentation of the main characteristics and advantages of the AF in the context of data assimilation for HdSs, see Section 2.2.

Assimilation experiments will be carried out with synthetic conventional data. The application of NNF produces a significant reduction in the computational effort. The goal of using the NN approach is to obtain a similar quality for analyses with better computational performance for prediction process.

The chapter is organized as follows. In the next section, a brief description of the NN and the AF for filtering problems is given. Section 3 presents the scheme of the NNF algorithm. The adaptation scheme for the NNFs is developed in Section 4. In

Section 5, the simulation experiments on filtering problems for moderate models are presented. Here first the one-dimensional system is presented to see clearly the idea and structure of the proposed NNAF. The assimilation experiment with the widely used Lorenz system, being of chaotic character, is also considered. The experiment with the ocean model MICOM (Miami Isopycnic Coordinate Ocean Model) for estimation of the oceanic current in different layers by the NNF and NNAF is described in Section 6, which demonstrates a possibility of the use of NNAF for state estimation in very high dimensional systems. It will be seen that the NNAF can improve considerably the performance of the NNF, which is beneficial for decision on future planning installation of stations exploiting renewable energy resources such as the tides and currents, as well as the waves and the thermal energy that resides in the seas. The conclusions are given in Section 7.

2. Artificial neural network filtering

2.1 Artificial neural network

Artificial neural networks (ANNs or NNs) are computing systems that are inspired by biological neural networks. Such systems are designed to "learn" by considering samples to perform the specified tasks. NNs are based on a collection of connected units or nodes called artificial neurons, which model the neurons in a biological brain. A standard simple NN scheme is shown in **Figure 1**. Each connection, like the synapses in a biological brain, can transmit a signal to other neurons. An artificial neuron that receives a signal then processes it and can signal neurons connected to it.

Neural networks have been extensively investigated in the context of adaptive control and system identification [15], but at the present, not so much in state estimation algorithms.

There are following principal types of NN: (i) feedforward neural network; (ii) radial basis function neural network; (iii) recurrent neural network (RNN); (iv) convolutional neural network (CNN); and (v) modular neural network. For more details, see Ref. [16]. For example, the RNN and CNN are designed to perform different tasks: the RNN is useful for *learning* the structure and syntax of the *language*, whereas the CNN is complex feedforward NN used for image classification and recognition because of its high accuracy, and it works by extracting features of image (identifying faces, individuals, street signs, and many other aspects of visual data).

NNs offer the useful properties such as linearity, nonlinearity, learning with teacher (modification of the NN weights), and adaptivity (capability to adapt their weights to changes in the surrounding environment). A neuron model scheme is given in **Figure 2**. One has

$$v_k = \sum_{i=1}^{m} W_{k,i} x_i + b_k, y_k = \varphi(v_k) \tag{1}$$

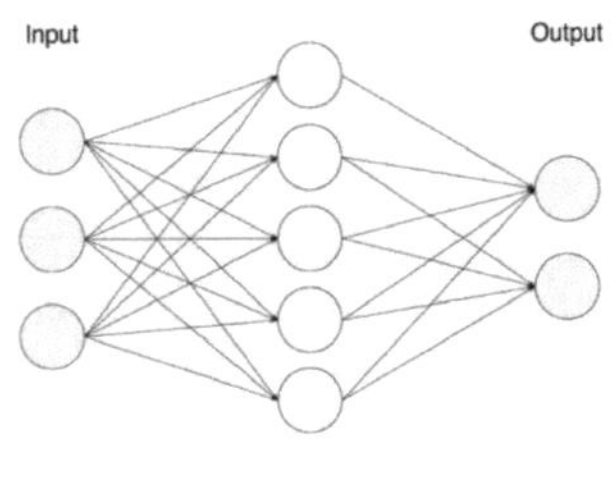

Figure 1.
Standard simple NN scheme: input, hidden, and output.

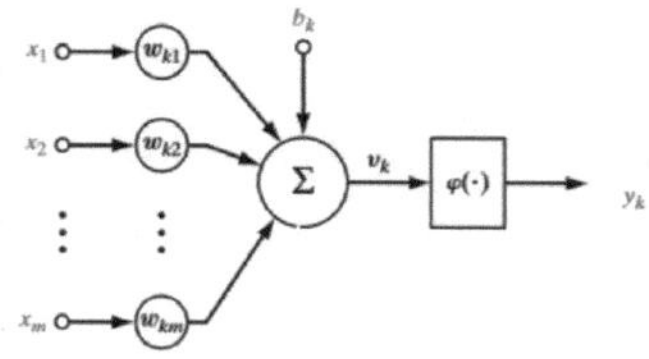

Figure 2.
Model of a neuron.

where $W_{k,i}$ is the synaptic weight between neuron i (at some l^{th} layer) and neuron k at the $(l+1)^{th}$, $x_1, \dots, x_n$ are the inputs, $\varphi(.)$ is the activation function, b_k is the bias, and y_k is the output signal of the neuron k at the $(l+1)^{th}$ layer.

There are different activation functions (AcF) such as Step, Linear, and Sigmoid functions. In **Figure 3**, three AcFs, Step, Linear, and Sigmoid, are displayed.

One of the most important results concerning a capacity of NN as a powerful algorithm for function approximation is expressed in terms of the **universal approximation theorem,** which states that a feedforward network with a single hidden layer containing a finite number of neurons can approximate continuous functions on compact subsets of R^n under mild assumptions on the AcF. The theorem thus states that simple neural networks can *represent* a wide variety of functions when given appropriate parameters. One of the first versions of the theorem was proved by Cybenko in 1989 for sigmoid activation functions [17].

In this chapter, the NN is designed in order to solve assimilation problems. The design procedure consists in the following steps:

Step 1. The choice of NN structure: What is the structure for the NN? This concerns the questions on (i) How many layers to be chosen? (ii) What are the connections between different nodes in different layers? (iii) What are the values to be initialized for the weights?

Step 2. The choice of variables for the input and output signals.

Step 3. Formulate an optimization problem for estimating the NN weights.

Step 4. Solve the formulated optimization problem: What is the optimization algorithm to be used?

All these questions will be addressed in details in Sections 3–6.

The objective of this chapter is to show that there exists a simple NNF structure, which can work efficiently and produce a good estimation results for the system states. As will be seen, all the NNFs employed in the experiments will have only one hidden layer. As to the AcFs, three types of AcF will be used: Linear, Sigmoid, and Step AcFs.

2.2 Adaptive filter

The AF for state estimation problem is defined as a filter, minimizing the MPE of the system output. One of the main features of the AF is that it is optimal among

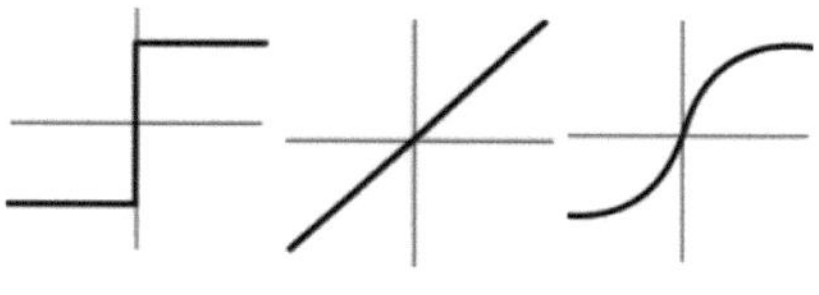

Figure 3.
Typical AcFs: step, linear, and sigmoid functions.

members of the set of parameterized state-space innovation representations, with the vector of pertinent parameters of the gain as a control vector. The assimilation problem considered in this chapter is formulated as a standard filtering problem

$$x(k+1) = \Phi x(k) + w(k) \tag{2}$$

$$z(k+1) = Hx(k+1) + v(k+1), k = 0, 1, \ldots \tag{3}$$

where $x(k)$ is the n-dimensional system state at the moment k, and $z(k)$ is the p-dimensional observation vector. Under standard conditions, related to the noise sequences $v(k), w(k)$, the optimal in mean squared estimate for $x(k)$ can be obtained from the system of recursive equations known as the KF.

$$\widehat{x}(k+1) = \Phi\widehat{x}(k) + K\varsigma(k+1), \varsigma(k+1) := z(k+1) - \Phi\widehat{x}(k) \tag{4}$$

where $K = K(M)$ is the gain matrix, which is a function of $M := M(k+1)$—the error covariance matrix (ECM) for the state prediction error (PE).

$$e(k+1) = \widehat{x}(k+1/k) - x(k+1), \widehat{x}(k+1/k) = \Phi\widehat{x}(k) \tag{5}$$

It is mentioned that in the KF, there exists a nonlinear matrix Riccati equation for computing $M(k+1)$. For a system with dimension of order10^7–10^8, under most favorable conditions (perfect knowledge of all system parameters and noise statistics), it is impossible to implement the KF (4) in the most powerful computers due to computational cost and insufficient memory.

In order to exploit the optimal structure of the filter (4), an AF is proposed in Ref. [13]. The filter is of the form (4) with the gain $K = K(\theta)$, which is a function of some vector θ of pertinent parameters. The choice of gain structure and its parameterization (i.e., θ) from the point of view of filter stability is studied in Ref. [18]. In Ref. [19], a simple and efficient algorithm is proposed for estimating the prediction error covariance matrix (ECM) based on the hypothesis on the separability of vertical and horizontal structure of the ECM and its parameterization. This allows to estimate optimally the unknown parameters of the structured parameterized ECM as well as to approach numerically the solution of the traditional Nearest Kronecker Problem in a simple and efficient way.

The optimality of the AF is understood in the Minimum Mean Squared (MMS) sense, which minimizes the objective function:

$$J(\theta) = E[\Psi(\varsigma)], \Psi(\varsigma) = \|\varsigma(k;\theta)\|^2$$
$$J = J(\theta) \rightarrow \min \arg\theta \tag{6}$$

with $E(.)$ is the mathematical expectation, and $\|.\|$ is the Euclidean norm. In Eq. (6), $\varsigma(k;\theta)$ is the PE for the system outputs, which depends on the vector of parameters θ. It is seen that if the KF minimizes the MMS *filtered* error for the system state (in state space), the AF minimizes the MMS *prediction* error for the system outputs in the observation space by tuning some pertinent parameters of the filter gain.

2.3 Simultaneous perturbation stochastic approximation (SPSA)

The choice of Eqs. (4) and (6) is important in many aspects in order to obtain a simple and efficient data assimilation algorithm. The idea lying behind Eqs. (4) and (6) is to select the vector $\theta \in \Theta$ consisting of some pertinent parameters as control variables for minimizing the cost function (6). Here Θ is the set of admissible values of θ, which ensure a filter stability.

The solution to the problem (6) can be found iteratively using a stochastic optimization (SA) algorithm.

$$\theta(k+1) = \theta(k) - \gamma(k+1)\nabla_\theta \Psi[\varsigma(k+1;\theta(k))] \tag{7}$$

where $\gamma(k)$ is a sequence of positive scalars satisfying some conditions to guarantee a convergence of the estimation procedure. The standard conditions are

$$\gamma(k)>0, \sum_{k=1}^{\infty}\gamma(k) = \infty, \sum_{k=1}^{\infty}\gamma^2(k) < \infty \tag{8}$$

The algorithm of Eqs. (4), (8), (11) is very simple to implement: at the kth assimilation instant, we need to integrate the numerical model Φ by $\hat{x}(k)$ to produce the forecast $\hat{x}(k+1/k)$ and to update $\theta(k)$ according to Eq. (7). The algorithm of Eq. (7) is completely defined if the gradient of the sample cost function $\nabla_\theta \Psi[\varsigma(k+1;\theta(k))]$ is available. This gradient can be computed using the adjoint equation (AE) approach [4]. However, writing the AE code is very complicated and hard task for complex physical systems. Moreover, it requires a linearization of the system (2) and (4).

A much less computational burden can be achieved by measuring the sample objective function (but not based on a gradient formula): one approximates the gradient using the values of the cost function [on the basis of finite difference scheme (FDSA)]. This can be done by applying a so-called component-wise derivative approximation algorithm. However, for high-dimensional systems, this algorithm is inapplicable: for approximation of each partial derivative, we need to make two integrations of the model (2).

Let us look at the class of SPSA algorithms [20, 21]. The algorithm SPSA is of the same structure as that of FDSA, with the difference that it perturbs not each component of θ separably, but perturbs stochastically and simultaneously all its components. Concretely, let $\Delta_k := (\Delta_{k,1}, \ldots, \Delta_{k,n})^T$ be a random vector, $\Delta_{k,i}, i = 1, .., n$ are Bernoulli independent identically distributed (iid). The gradient of the sample objective function is estimated as

$$g = \nabla_\theta \Psi = (g_1, \ldots, g_n)^T, \tag{9}$$

$$g_i = [\Psi(\theta(k) + c_k\Delta_k) - \Psi(\theta(k) - c_k\Delta_k)]/(2c_k\Delta_{k,i}), i = 1, \ldots, n. \tag{10}$$

One sees from Eqs. (9) and (10) that all the directions in the parameter space are perturbed at the same time (the numerator is identical in all components). Thus, SPSA uses only two (or three) times the direct integration of the model, independently on the dimension of θ, which makes it possible to apply to high dimensional optimization problems. No development of the AE code is needed. Generally, SPSA converges in the same number of iterations as the FDSA, and it follows approximately the steepest descent direction, behaving like the gradient method [20, 21]. In fact, SPSA, with a random search direction, does not follow exactly the gradient path. In average, though, it tracks the gradient nearly because the gradient approximation is almost an unbiased estimator of the gradient, as shown in Refs. [20, 21]. For SPSA algorithm, the conditions for the sequence of positive c_k are

$$c_k>0, \sum_{k=1}^{\infty} c_k = \infty, \sum_{k=1}^{\infty}(\gamma(k)/c_k)^2 < \infty \tag{11}$$

where $\gamma(k)$ satisfies (8).

3. NN filtering (NNF)

Let us return to the input–output system (2)–(3). Instead of the AF, we will construct a NNF for estimating the system state $x(k)$ based on the set of observations from beginning up to the moment k. According to the NN scheme, we have to define the set of inputs and the set of output signals. These two sets are introduced as

$$S_I := \{\hat{x}_{nn}(k/k-1), z(k)\}, S_O := \{\hat{x}_{nn}(k)\} \tag{12}$$

where S_I is the set of inputs, and S_O is the set of output signals, and the estimate $\hat{x}_{nn}(k/k-1) = \Phi\hat{x}_{nn}(k-1)$ is the forecast obtained from the integration of $\hat{x}_{nn}(k-1)$ by the numerical model Φ. It is mentioned that $\hat{x}_{nn}(k-1)$ is the filtered estimate produced by the NNF, and symbolically, it is obtained by application of the NNF on the basis of the inputs S_I subject to the weights $W(k)$,

$$\hat{x}_{nn}(k) = f_{nn}[\hat{x}_{nn}(k/k-2), z(k); W(k)] \tag{13}$$

The difference between $\hat{x}_{nn}(k)$ in Eq. (13) and $\hat{x}(k)$ in Eq. (4) is that $\hat{x}_{nn}(k)$ is the estimate produced by the NNF, which is of the NN structure. Let us introduce a new model

$$x_m(k+1) = \Phi x_m(k) + w_m(k) \tag{14}$$

with $w_m(k)$ being a stochastic process, which may be different from $w(k)$. The reason is that in practice we are given not exactly the information on the process $w(k)$ (in the experiments for simplicity, we assume $w_m(k) = 0$). The vector $x_m(k)$ will serve as samples for learning process to estimate the NN parameters W. For illustration, one scheme of NNF structure for the system with two-dimensional state and two-dimensional observation vector is given in **Figure 4**.

In **Figure 4**, $\hat{W}_l(i,j)$ denotes the weight between the i node at the lth layer and the j node at the next layer. The algorithm for updating the weights in the NNF looks as follows:

Algorithm 3.1 (Learning algorithm for estimating NNF weights).

Suppose the set of observations $z(k), k = 1, \ldots, N$ is available. At the moment $k = 1$, let the a priori values $\hat{W}_l(k-1; i,j)$, $\hat{x}_{nn}(k/k-1)$ be given. Suppose we are given the set of observations $z(k), k = 1, \ldots, N$.

Step 1. Pass the inputs $\hat{x}_{nn}(k/k-1)$, $z(k)$ through the NNF. The NNF produces the output (estimate) $\hat{x}_{nn}(k)$.

Step 2. Compute the state error $e_{nn}(k) = \hat{x}_m(k) - \hat{x}_{nn}(k)$.

Step 3. Using the SPSA algorithm to update the NNF weights according to

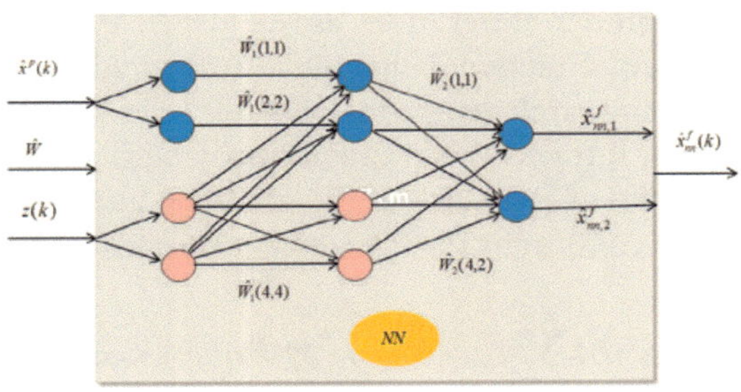

Figure 4.
Scheme of NNF structure.

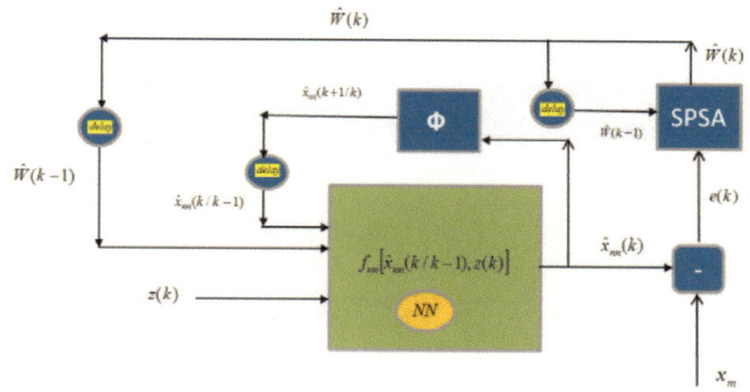

Figure 5.
Scheme for Algorithm 3.1.

$$\hat{W}(k) = \hat{W}(k-1) - \gamma(k)\nabla_W \Psi\left[\hat{W}(k-1)\right] \tag{15}$$

by minimizing the cost function

$$J_{nnf}(W) := E[\Psi(e_{nn}(W))] = E\left[\|e_{nn}(k;W)\|^2\right] \rightarrow \min \arg(W) \tag{16}$$

Step 4. If $k < N$, go to Step 1. Otherwise, go to Step 5.
Step 5. Stop.
The scheme for *Algorithm 3.1* is shown in **Figure 5**.

Comment 3.1. The sample gradient $\nabla_W \Psi\left[\hat{W}(k-1)\right]$ is evaluated as shown in Section 2.3 with $\gamma(k)>0$ (jointly with $c_k>0$) satisfying two conditions (8) and (11) ensuring a convergence of the algorithm (15) [20]. The sequence $\{c_k\}$ of positive scalar values participates in generating simultaneous perturbations to approximate $\nabla_W \Psi\left[\hat{W}(k-1)\right]$.

Comment 3.2. The estimates for the system state in (2) are $\hat{x}_{nn}(k)$, which are produced during running of *Algorithm 3.1*. To obtain the better estimates on the basis of nonadaptive NNF, the NNF should be rerun subject to the weights resulting from applying *Algorithm 3.1*.

Comment 3.3. If the number of observations is insufficient for well estimating the NNF coefficients, it is possible to simulate a longer sequence of pseudo-observations

$$z_m(k) = Hx_m(k) + v_m(k)$$

where $v_m(k)$ may be different from $v(k)$ (but chosen as closely as possible to $v(k)$ in the statistical sense). The similar experiment will be carried out in this framework for the ocean model MICOM in Section 6.

4. Adaptive NNF (NNAF)

To improve the performance of the NNF, the NNAF is developed in this section. The NNAF is of the same structure as that of the NNF, but it is designed to minimize the MPE of the system output. The NNAF is based on the second optimization procedure with the initial weights resulting from application of *Algorithm 3.1*.

For the design of the NNAF, the following objective function is introduced

$$J_{nnaf}(\theta) := E\left[\Psi(\varsigma_{nn}(W(\theta))\right] = E\left[\|\varsigma_{nn}(k;W(\theta))\|^2\right] \rightarrow \min \arg(\theta) \tag{17}$$

The difference between two objective functions (16) and (17) is that if (16) is expressed in terms of the (filtered) state estimation error, the error function (17) measures the difference between the observation and the prediction of the system output. As to the parameters to be adjusted, the following parameterizations are possible to be used:

$$W(\theta) = \hat{W}_o + \theta \tag{18}$$

$$W(\theta) = \theta\hat{W}_o \tag{19}$$

where $\hat{W}_o$ is the estimate obtained from the learning procedure (*Algorithm 3.1*). The initial value for θ is $\theta(0) = 0$ for the parameterization (18) and $\theta(0) = 1$ for (19). It is mentioned that when the structure of the NNAF is closed to the AF in Ref. [19], the parameterization (19) is more appropriate. We have now

$$\begin{gathered} \varsigma_{nn}(k; W(\theta)) = z(k) - H\Phi\hat{x}_{nn}(k-1) \\ \hat{x}_{nn}(k-1) = f_{nn}[\hat{x}_{nn}(k-1/k-2), z(k-1); W(\theta)] \end{gathered} \tag{20}$$

More concretely, the NNAF is a result of the following.

Algorithm 4.1 (Adaptive procedure for improving NNF performance).

At the initial moment $k = 1$ suppose we are given $\hat{x}_{nn}(k/k-1)$ and $\hat{W}_l(i,j)$ ($\hat{W}_l(i,j)$ are obtained at the end of applying *Algorithm 3.1*). Suppose the set of observations $z(k), k = 1, \ldots, N$ is available.

Step 1. Pass the inputs $\hat{x}_{nn}(k/k-1), z(k)$ through the NNF initialized by the weights $\hat{W}_l(i,j)$. The NNF produces in the output the estimate $\hat{x}_{nn}(k)$.

Step 2. Compute the error $\varsigma(k) = z(k) - H\hat{x}_{nn}(k/k-1)$ where.

$$\hat{x}_{nn}(k/k-1) = \Phi\hat{x}_{nn}(k-1), \hat{x}_{nn}(k-1) = f_{nn}[\hat{x}_{nn}(k-1/k-2), z(k-1); \hat{W}_l(i,j)].$$

Step 3. Using the SPSA algorithm to update the weights

$$\theta(k) = \theta(k-1) - \gamma(k)\nabla_{\theta(k-1)}\Psi[\theta(k-1)] \tag{21}$$

for seeking the optimal parameters θ to minimize the cost function

$$J_{nnf}(\theta) := E\left[\Psi(\varsigma_{nn}(k; \hat{W}(\theta)))\right] = E\left[\left\|\varsigma_{nn}(k; \hat{W}(\theta))\right\|^2\right] \rightarrow \min \arg\theta \tag{22}$$

Step 4. If $k < N$, go to Step 1. Otherwise, go to Step 5.

Step 5. Stop.

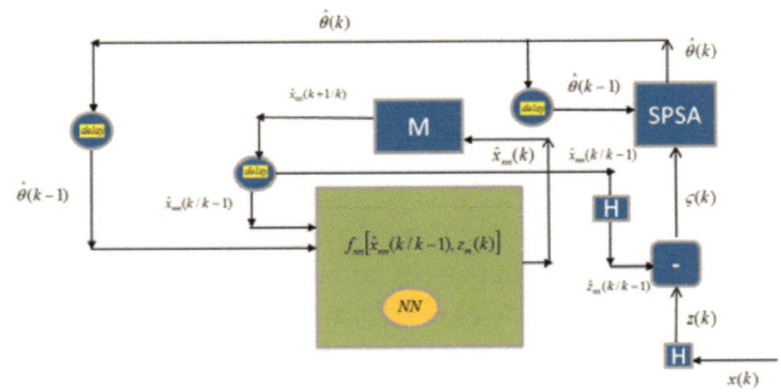

Figure 6.
Scheme for Algorithm 4.1.

Figure 6 displays the scheme of *Algorithm 4.1*. Compared to **Figure 3**, here the weights are assumed given (as resulting from application of *Algorithm 3.1*). States of the numerical model are no longer participating in the SPSA algorithm; instead, the innovation vector is involved in the optimization process.

5. Numerical experiments for systems of moderate dimensions

5.1 Experiment 1: One-dimensional system

We present in this section a very simple numerical experiment with an one-dimensional system to illustrate clearly the idea on the NNAF.

Let us consider the system of Eqs. (2)–(3) subject to $\Phi = 0.99, H = 1, Q = 0.09, R = 0.01$. The experiment is carried out on the basis of a NNF, and the performance of the NNF is compared with that of the KF and NNAF. The structure of the NNF is shown in **Figure 7**.

Thus, the NNF has one hidden layer structure. The Linear and Sigmoid AcFs are used in this NNF. The weights $W_l(i,j)$ are estimated by applying *Algorithm 3.1*. Here the error $e(k) = x(k) - x_m(k)$ is the difference the "wrong" model state governed by $x_m(k+1) = \Phi x_m(k), k = 0, 1, \ldots, N$ and the true system state governed by $x(k+1) = \Phi x(k) + w(k), k = 0, 1, \ldots, N$. The performance of the NNF is expressed in the terms of root mean square (RMS) of the state prediction error.

Figure 8 shows the performances of three filters—NNF, NNF-FIX, and NNAF. The red curve "*NNF*" is the RMS of state prediction error produced by the NNF during the optimization process (*Algorithm 3.1*). The green curve "*NFF-FIX*" is produced by the NNF with the fix weights obtained at the end of *Algorithm 3.1*. As to the blue curve "*NNAF,*" it displays the performance of the NNAF by applying *Algorithm 4.1*. The reason for which the NNAF has produced the best performance, compared to the NNF and NNF-FIX, is that the NNAF minimizes the MPE for the system outputs. It means that learning process does not extract all useful information on the system state contained in the observations.

Time evolution of the weights, associated with connecting links between the nodes in the hidden layer and the node of the output layer, is displayed in **Figure 9**.

Figure 10 shows the performances of four filters implemented for estimating the system state of the 1d model: NAF (nonadaptive filter), AF developed in Ref. [18], NNF (*Algorithm 3.1*), and NNAF (*Algorithm 4.1*). The NAF is the filter with the constant gain computed on the basis of the stationary solution of the Riccati equation associated with the KF. One sees that the NNF behaves better than the NAF at the beginning of the assimilation process. Application of adaptation allows to

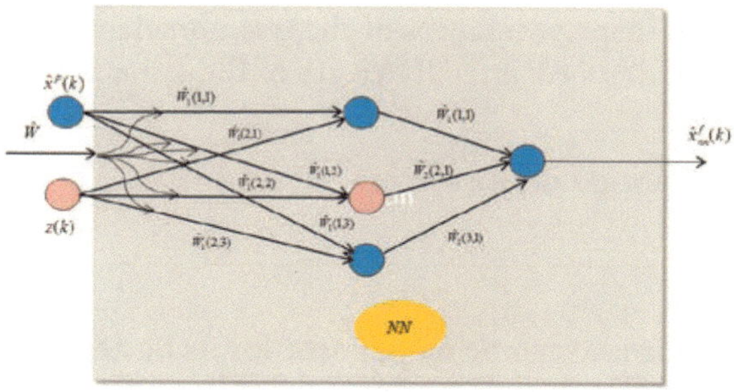

Figure 7.
Structure of the NNF for the one-dimensional system.

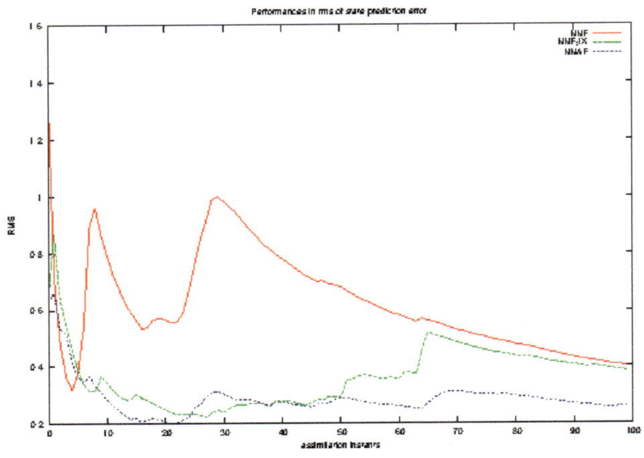

Figure 8.
*RMSs of state prediction error resulting from three filters: NNF (*Algorithm 3.1*, red line), NNF-FIX (green line), and NNAF (*Algorithm 4.1*, blue line). The error produced by the NNF during the optimization process is too high compared to that of the NNF-FIX. It (NNF) reaches the same level (of NNF-FIX) at the end of the assimilation period. As to the NNAF, the error remains low during all assimilation periods (after about the first 10 iterations).*

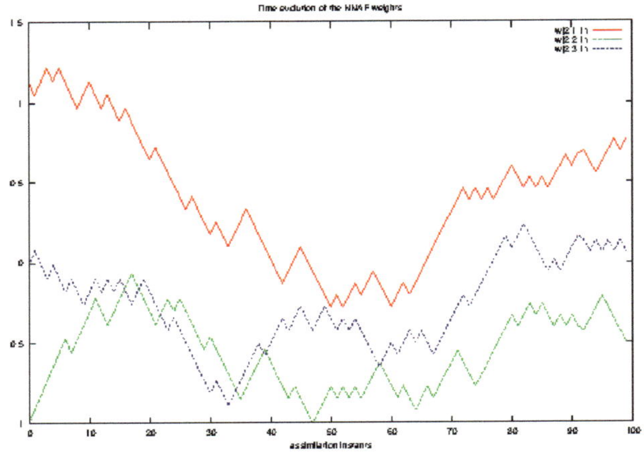

Figure 9.
Time evolution of several weights (between the nodes in the second (hidden) layer and the nodes in the output layer) in the NNAF during the adaptation process: the weight of the connection link between the first node (hidden layer) and the output node (red curve); the weight of the connection link between the second node (hidden layer) and the output node (green curve); and the weight of the connection link between the third node (hidden layer) and the output node (red curve).

improve considerably the performances of the two nonadaptive NAF and NNF. In general, two adaptive filters AF and NNAF are of the same performance.

5.2 Experiment 2: Chaotic Lorenz system

5.2.1 Lorenz equations

The Lorenz attractor is a chaotic map, noted for its butterfly shape. The map shows how the state of a dynamical system evolves over time in a complex, non-repeating pattern.

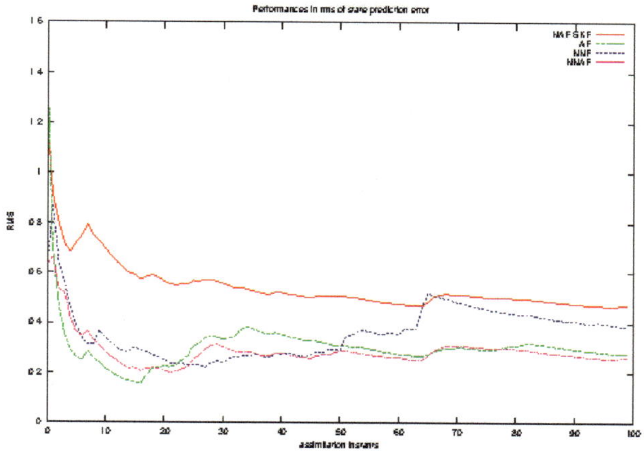

Figure 10.
*RMSs of state prediction error resulting from four filters: NAF (nonadaptive filter, red line), AF (green line), NNF (blue line), and NNAF (pink line). The NNF, thanks to the learning process, behaves better than NAF. The AF developed in Ref. [18] and NNAF (*Algorithm 4.1*) are of the same performance.*

The attractor itself and the equations from which it is derived were introduced by Lorenz [22], who derived it from the simplified equations of convection rolls arising in the equations of the atmosphere.

The equations that govern the Lorenz attractor are:

$$dy_1/dt = -\sigma(y_1 - y_2),$$
$$dy_2/dt = \rho y_1 - y_2 - y_1 y_3, dy_3/dt = y_1 y_2 - \beta y_3 \quad (23)$$

where σ is called the Prandtl number, and ρ is called the Rayleigh number. All $\sigma, \beta, \rho > 0$, and usually $\sigma = 10, \beta = 8/3, \rho$ are varied. The system exhibits chaotic behavior for $\rho = 28$ but displays knotted periodic orbits for other values of ρ.

5.2.2 Numerical model

In the experiments, the parameters σ, β, ρ are chosen to have the values 10, 28, and 8/3, respectively, for which the "butterfly" attractor exists.

The numerical model (NM) is obtained by applying the Euler method (first-order accurate method) to approximate Eq. (23). Symbolically, we have the NM

$$y(t_{k+1}) = F[y(t_k)], y(t_{k+1}) := [y_1(t_{k+1}), y_2(t_{k+1}), y_3(t_{k+1})]^T \quad (24)$$

where $\delta t := t_{k+1} - t_k$ is the *model* time step. The observations arrive at the assimilation instants T_k and $\Delta T_k := T_{k+1} - T_k$.

5.2.3 Numerical model and observation

The numerical model is derived subject to $\delta t := 0.005$. The true states $x(k)$, produced by the NM corrupted by additive noise at the moment T_k subject to $\Delta T_k = 25\delta t$, are taken as observation $z(k), k = 1, \dots, N_o$ contaminated with the noise v_k having zero mean and variance σ_k^2, with $\sigma_k = 0.1$. The true dynamical system, corresponding to the transition of the states between two assimilation instants T_k and T_{k+1}, is expressed as

$$x(k+1) = F[x(k)] + w(k) \tag{25}$$

In Eq. (25), $w(k)$ simulates the model error. The sequence $\{w(k)\}$ is assumed to be a white noise having diagonal covariance matrix with variances 2, 12.13, and 12.13, respectively. The observation system is given by

$$z(k+1) = Hx(k+1) + v(k) \tag{26}$$

where the observation operator $H = \left[h_1^T, h_2^T\right]^T, h_1 = (1, 0, 0), h_2 = (0, 0, 1)$, that is, the first and third components $x_1(k), x_3(k)$ are observed at each time instant k. The noise sequence $\{v_k\}$ is white with zero mean and variance $R = 2I_2$, where I_n is the unit matrix of dimension n. The initial estimate in all the filters is given by $\hat{x}(0) = (5, 10, 27)^T$.

The true model (TM) state $x^*(0)$ is modeled as the solution of Eq. (25) subject to the initial state $x^*(0) = (1.509, 1.531, 25.461)^{\mathrm{T}}$.

The problem in this experiment is to apply the NNF and its adaptive version to estimate the true system state using the observations $z(k)$, $k = 1, \ldots, N_o$, and to compare their performances with those produced by the AF.

5.2.4 NNF structure and numerical results

The NNF used in this section has the following structure.

The NNF has only one hidden layer, and its AcFs are chosen as Linear and Step AcFs (**Figure 3**). The structure in **Figure 11** is chosen to be as simple as possible. It is interesting to ask whether there are more complicated NNFs, which can yield a better performance? However, in this chapter, we restrict our self to considering only the NNF with the structure in **Figure 11**.

The weights in the NNF are first estimated (learning process) using the sequence of the erroneous model (EM) states [which is a noise-free version of the TM (true model) (25)]. The initial state in the EM is the same as that in the TM. Typical evolution of weights during the learning process is shown in **Figure 12**. It is seen that the weights' estimates are relatively stable, though we cannot say they are already stabilized due to an insufficient number of iterations. All the weights are initialized with the values 0.5. We have also tested the NNF for the situation with

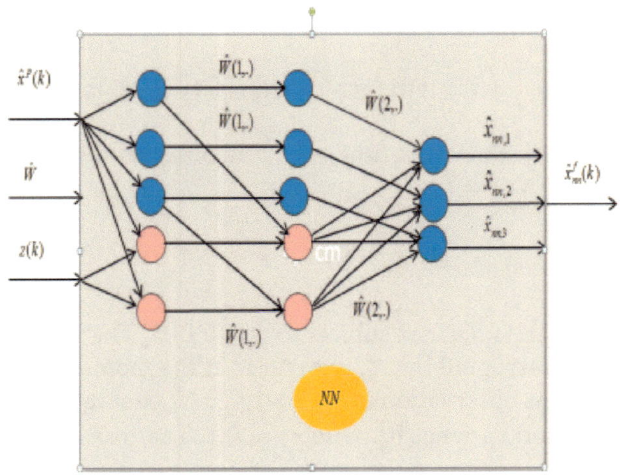

Figure 11.
Structure of the NNF for the Lorenz data assimilation system.

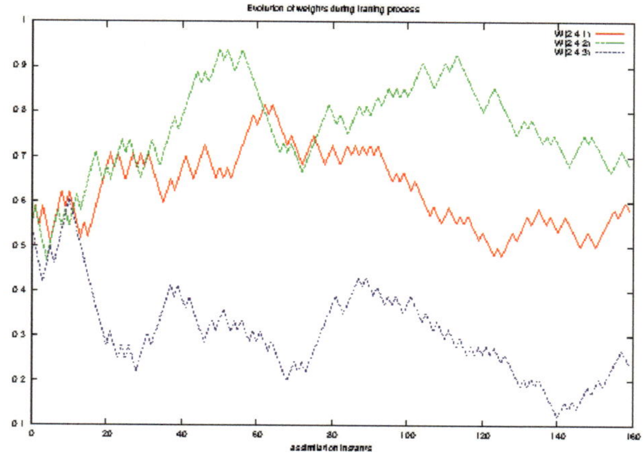

Figure 12.
Typical time evolution of NN weights (between the nodes in the second (hidden) layer and the node in the output layer) in the NNAF during the learning process: the weight of the connection link between the fourth node (hidden layer) and the first node of the output layer (red curve); the weight of the connection link between the fourth node (hidden layer) and the second node of the output layer (green curve); and the weight of the connection link between the fourth node (hidden layer) and the third node of the output layer (red curve). One notes that the weights do not yet attend a stationary regime.

the initial weight values equal to 1. The results show that there is no significant difference in filter performance between these two situations.

It is mentioned that the Lorenz system is very sensitive to the initial state condition. Different initial states will lead to a variety of solution behavior: some solutions are steady; others oscillate between two or more states; and still others vary in an irregular manner.

We generate observations at 160 assimilation instants (4000 model time steps). The different filters will be implemented to assimilate the observations.

Figure 13 shows the performance of three filters—NFF1, NNF2, and NNF3, which are initialized with the same initial condition $\hat{x}(0) = (5, 10, 27)^T$. It is seen

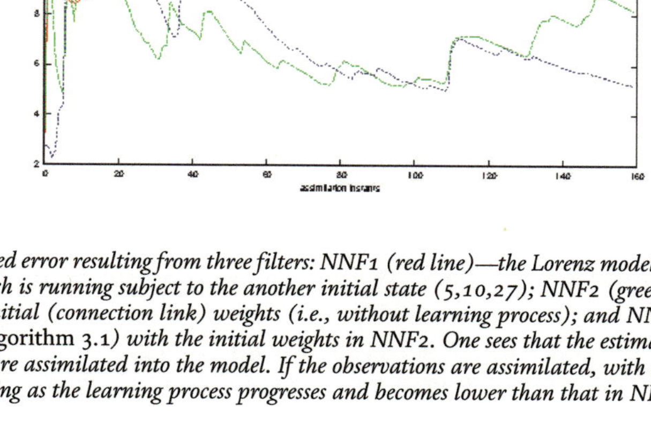

Figure 13.
RMSs of state filtered error resulting from three filters: NNF1 (red line)—the Lorenz model alone (without data assimilation), which is running subject to the another initial state (5,10,27); NNF2 (green line)—the NNF running with fix initial (connection link) weights (i.e., without learning process); and NNF3 (blue line)—the NNF (applying Algorithm 3.1*) with the initial weights in NNF2. One sees that the estimation error is too high if no observations are assimilated into the model. If the observations are assimilated, with the learning process, the error is decreasing as the learning process progresses and becomes lower than that in NNF2 at the end of the learning process.*

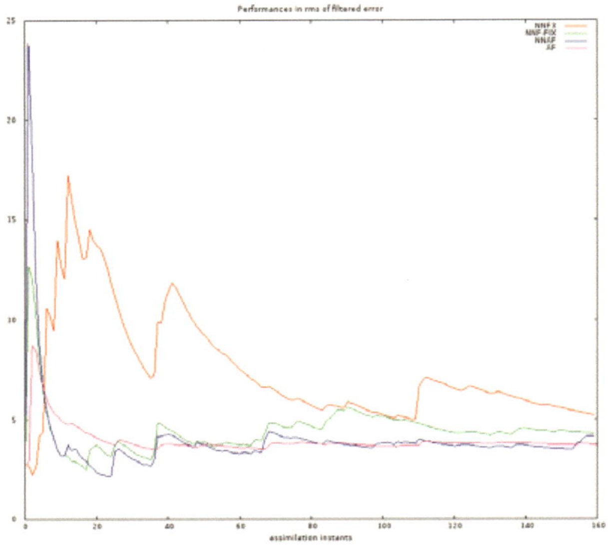

Figure 14.
RMSs of state filtered error resulting from three filters: NNF3 (red line)—the same NNF3 in ***Figure 13*** (Algorithm 3.1)*; NNF-FIX (green line)—the NNF subject to the fix coefficients obtained at the end of the optimization process in NNF3; and NNAF (blue line)—the NNAF, which is obtained by tuning the weights in NNF (initialized by the values obtained at the end of NNF3) to minimize the prediction error for the system outputs* (Algorithm 4.1) *AF (pink line)—the AF (4) and (6) developed in Ref. [19].*

that without data assimilation, the erroneous initial condition leads to big errors between the true states and the states of the erroneous model (red curve, NNF1). By simply initializing the weights in the NNF, it is possible to reduce an important portion of the errors (green curve, NNF2). At the beginning of the learning process (NNF3), the error is relatively high (even higher than that in NNF2), but it quickly decreases after about 20 assimilation instants.

The SPSA algorithm is really capable of well searching optimal weights as seen from comparison of performances of the two filters NNF3 and NNF4 (**Figure 14**). Finally, tuning the weights in the NNF to minimize the prediction error (NNAF) of the system output yields the best results (among all the filters of the neural network structure) as seen from the curve "*NNAF*" (*Algorithm 4.1*). To verify the optimality of the NNAF, we have also implemented the AF (4) and (5) based on tuning the parameters of the filter gain [19]. The performance of the AF is shown by the pink curve "*AF*" in **Figure 14**, which is almost the same as that of the NNAF. As before, all the filters are starting from the same initial condition as used in the NNF1. It is interesting to note that the two adaptive filters NNAF and AF are of different structures, but they can yield the same performance. If for the NNAF, the choice of the NN structure (sets of inputs, outputs; number of layers, initial values of the weights to be selected, and learning process) is most important, in the AF the question about the choice of stabilizing gain with appropriate parameterization is primordial for the success of the AF.

6. Numerical experiment for high dimensional ocean model MICOM

The oceans represent a vast source of renewable energy. In general, ocean energy can be divided into six types of different origin and characteristics: ocean wave, tidal range, tidal current, ocean current, thermal energy, and salinity gradient. Knowledge of these quantities is important for decision making on sizing, placement, and optimization of installation of system energy.

These oceanic quantities are now can be well modeled through an ocean numerical model, which is constructed according to the ocean configuration of interest. The difficulties arising here are that the model solution gives us only the relative values of quantities of interest because the model is always far from the reality. Therefore, in order to more accurately assess the true values of oceanic quantities, the measurement data are required to be inserted into the model to correct the model solution. The role of DA algorithms is just to properly (in some sense, in the best way) insert data into a numerical model so that to ensure a stability of the algorithm and to minimize the estimation error.

In this section, the experiment with the ocean model MICOM is carried out, which is aimed at estimating the ocean current in different depths (vertical layers).

6.1 Numerical model

The ocean model used for the DA experiment with the NNF and NNAF is the MICOM (Miami Isopycnic Coordinate Ocean Model), which is identical to that described in Ref. [18]. The model configuration is a domain situated in the North Atlantic from 30°N to 60°N and 80°W to 44°; for the exact model domain and some main features of the ocean current produced by the model, see Ref. [19].

The MICOM used here has the state $x = (h, u, v)$, where $h = h(i, j, k)$ is a layer thickness, and $u = u(i, j, k)$ and $v = v(i, j, k)$ are two velocity components. In the model, $i = 1, \ldots, 140; j = 1, \ldots, 180$; and $N_z = 4$ vertical layers, resulting in the state vector of dimension 302,400. The model is integrated from the state of rest during 20 years (*ys*). During the period of 2 years 19 and 20, every 10 days, the sea surface height (SSH) is computed as a linear function of the layer thickness *h*, which is considered as observation in assimilation experiments (totally 72 observations are available). To be closer to realistic situations with satellite observations, available only at along-track grid points, we assume that the observations are noise-free sea surface height (SSH), available only at the set $S_o$of horizontal grid points $S_o = \{(i, j), i = 1, 11, \ldots, 131, j = 1, 11, \ldots, 171\}$(but not at all horizontal grid points).

6.2 Structure of NNF

We first construct a NNF block for estimating the layer thickness variable *h* using the SSH observations. The Linear and Step AcFs are used in the NNF. The two velocity components $(u, v)$are computed using the geostrophy hypothesis. The scheme of the NNF is shown in **Figure 15**.

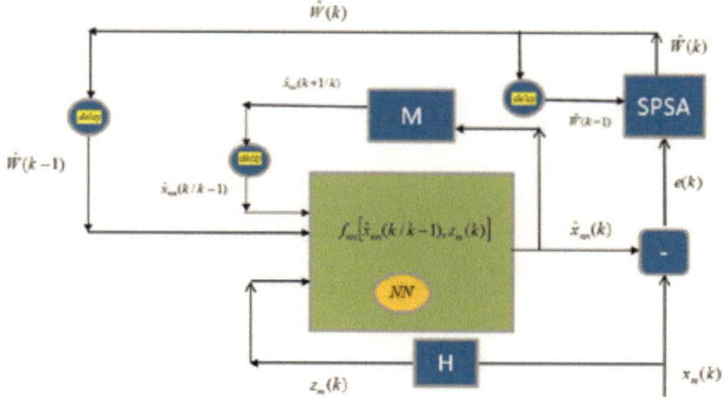

Figure 15.
Scheme for NNF for data assimilation in MICOM.

It is mentioned that as the observations are available only at the set S_o, one sub-NN block (with sigmoid AcF), named "optimal interpolation," is designed to interpolate the observations (i.e., to estimate the values of observations), available at S_o, on all the horizontal grid points of the surface. This operation is very important to ensure smooth corrections to all layer thicknesses and to estimate all layer velocities.

The NNF has one hidden layer as that used in the Lorenz experiment. One note from **Figure 15** that in the MICOM experiment, the *Algorithm 3.1* has been implemented not on the basis of the true observations $z(k)$ in Eq. (3) but using the other set of "observations" (calculated from $x_m(k)$, see **Figure 15**) for estimating the weights in the NNF (see *Comment 3.3*). To test the influence of "true samples" and "erroneous samples" on the NNF performance, we have implemented two learning procedures, one is based on the set of erroneous samples $x_m \propto x_{err}$, and another on the basis of true samples $x_m \propto x^*$.

6.3 Numerical results

In **Figure 16**, we show the evolution of some NNF weights during the learning process on the basis of the set of erroneous samples.

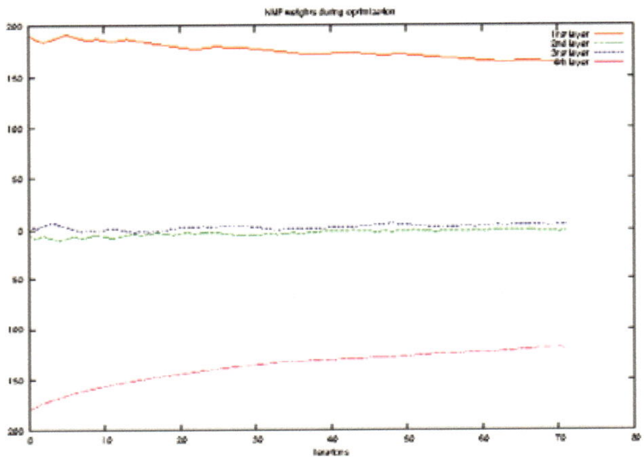

Figure 16.
Typical evolution of estimated weights in different layers during learning process. Initial weights are taken from application of the Cooper and Haines's water column lifting–lowering method [19].

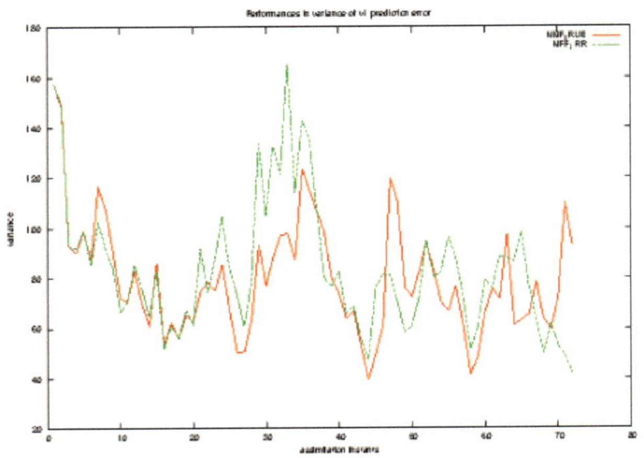

Figure 17.
Variances of prediction error for the velocity (v component) at the first layer by NNFT (red curve) and NNFE (green curve).

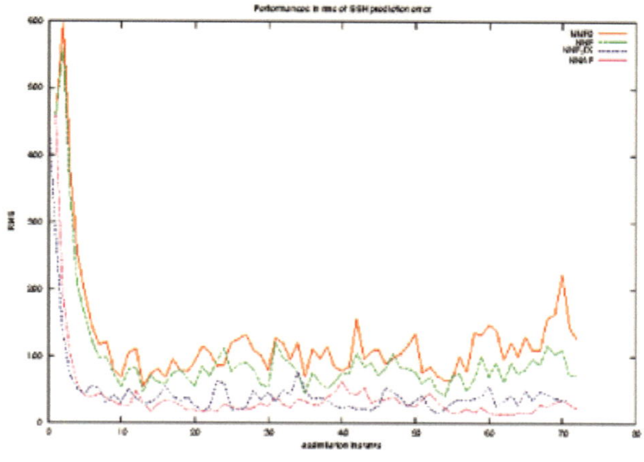

Figure 18.
Performance of different filters (in variance) of SSH prediction error (objective function).

The weights, obtained at the end of learning process, are inserted into two NNFs—NNFT and NNFE, where the weights in the NNFT are obtained by minimizing the difference between the NNF state and the true system state, and NNFE—with the weights resulting from minimizing the difference between the NNF state and the state of the erroneous model. **Figure 17** displays the variances of the prediction error for velocity (v component) at the first layer produced by NNFT (red curve) and NNFE (green curve). One sees that learning on the basis of samples of the true state yields more optimal weights, especially in avoiding the error pick, but generally speaking, only a slight improvement is observed.

We have tested different filters to see the effect of optimization in finding the weights in the NNF and those in the NNAF. The performances of these filters are shown in **Figure 18**. From **Figure 18**, the error in the *NNF*0 on the basis of the initialized weights (red curve) is too high compared to those of the other filters. The improvement is observed for the *NNF* (*Algorithm 3.1*) during the optimization process (green curve). If we run the *NNF* – *FIX* with the fix weights, resulting at the end of the optimization process in the*NNF*, a much better performance has been produced (blue curve). Finally, the best results are obtained by running the NNAF (*Algorithm 4.1*) (pink curve).

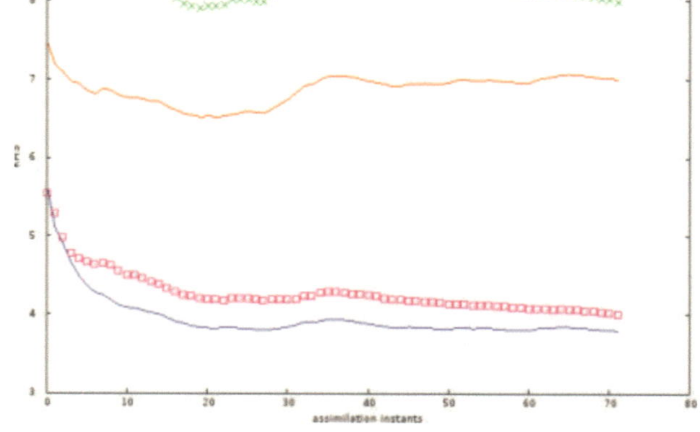

Figure 19.
Spatial temporal average RMS of velocity at first two layers for u and v component (by NNF).

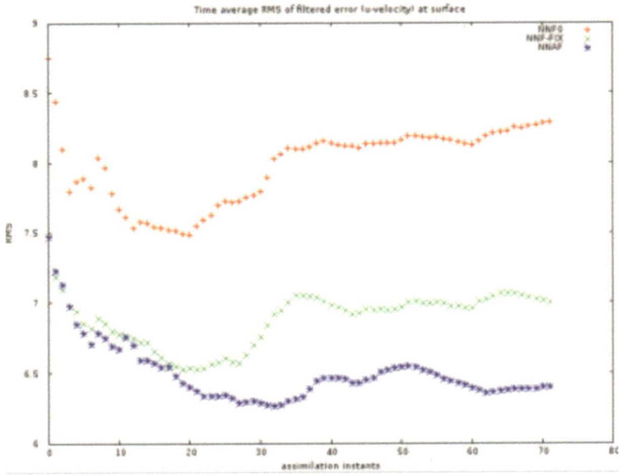

Figure 20.
Spatial temporal average RMS of u *velocity at surface produced by NNF, NNF-FIX, and NNAF.*

To have a more complete idea on how work the *NNF* and *NNAF* in high dimensional setting, in **Figure 19**, we show the spatial temporal average RMS of velocity (u,v) at the first two layers. As expected, estimating the velocity at the surface is more difficult than in deeper layers. One remark is that the u-velocity component is easier to estimate (i.e. with less error) than the v component. **Figure 20** gives a picture on performance of different filters. It displays the error (for u velocity at the surface) produced by three filters: $NNF0$, $NNF - FIX$, and $NNAF$. If in **Figure 18**, the difference (in *SSH* forecast error) between $NNF - FIX$ and *NNAF* is less visible for the velocity variable, and the superiority of the*NNAF* over the $NNF - FIX$ is remarkable. This tendency is observed in the comparison of velocity estimation errors produced by the three mentioned filters, at different layers as well as for the v-velocity component.

7. Conclusions

In this chapter, a constructive algorithm offering an attractive approach for the design of a NNAF is proposed. In particular, a simple NN structure with one hidden layer has proved to be efficient: it can yield a high performance of all the NNAFs implemented in the experiments—either with moderate or with high dimensional systems. The superiority (in performance) of the NNAF is confirmed even for the Lorenz system with chaotic character, which is very popular due to its high sensitivity to the initial condition. It would emphasize here that in the NNF and NNAF, no initial condition is used as a control variable to formulate the optimization problems. The experiment with the HdS ocean model MICOM demonstrates a possibility to exploit the NNF for state estimation problems for realistic HdSs.

It is found from the experiments that an appropriate choice of initial values for the weights is important for yielding a high performance of the NNF, with a moderate number of hidden layers and iteration steps. In the MICOM experiment, we have initialized the weights with the values corresponding to the lifting-lowering water column method. A more efficient way to initialize the weights (to achieve a better performance) is to follow the method described in Ref. [23] based on computing the dominant Schur vectors of the system dynamics. It is highly probable that the NNF with a larger number of hidden layers and nodes could produce a better performance. The question on how many numbers of layers and nodes to be chosen for the NNAF is open and is left for the future study.

The experimental results confirm that including adaptation as a mechanism for minimizing the MPE of the system outputs is an efficient way to improve the NNF performance.

We want to stress that the success of the NNAF is possible partly due to application of SPSA optimization algorithms. These algorithms are relatively ease of use for solving difficult multivariate optimization problems and are simple to implement and possess a good convergence rate. They require only two objective function measurements per iteration regardless of the dimension of the optimization problem. These features make SPSA as a powerful tool for solving optimization problems with nonlinearities and high dimensionality of complex dynamical systems.

Finally, as the developed NNAF in this chapter is very effective and easy to implement, which makes it quite applicable for solving different real-world tasks, in particular for solving DA problems in more sophisticated ocean models like the HYCOM (HYdrodynamic Coordinates Ocean Modeling). With the HYCOM model, it is possible to estimate, for example, thermal energy and salinity gradients—the quantities, serving important sources of renewable energy to generate electricity for global electrical demand in future. It is mentioned that the model HYCOM is used by the French Navy Hydrographic and Oceanographic Service (SHOM) for the prediction of ocean currents, to predict variations on the climatic scale, with a model of waves and a metric resolution, with very short-lived physics at the littoral scale. Implementation of the NNAF represents real challenges for the future operational forecasting systems [24].

Author details

Hong Son Hoang* and Remy Baraille
REC/HOM/SHOM, Toulouse, France

*Address all correspondence to: hhoang@shom.fr

References

[1] Kolmogorov AN. Stationary sequences in Hilbert space. Bulletin of Moscow University. 1941;**2**(6):1-40. In Russian

[2] Wiener N. Extrapolation, Interpolation, and Smoothing of Stationary Time Series With Engineering Applications. Cambridge: MIT Press; 1949

[3] Kalman RE. A new approach to linear filtering and prediction problems. Transaction of the ASME—Journal of Basic Engineering. 1960;**82**(Series D): 35-45

[4] Talagrand O, Courtier P. Variational assimilation of meteorological observations with the adjoint vorticity equation. I: Theory. Quarterly Journal of the Royal Meteorological Society. 1987; **113**:1311-1328

[5] Evensen G. Data Assimilation: The Ensemble Kalman Filter. Berlin: Springer; 2007

[6] Gordon NJ, Salmond DJ, Smith AFM. Novel approach to nonlinear/non-Gaussian Bayesian state estimation. IEE Proceedings of Radar and Signal Processing. 1993;**140**(2):107-113. ISSN 0956-375X

[7] Carrassi A, Bocquet M, Bertino L, Evensen G. Data assimilation in the geosciences: An overview on methods, issues and perspectives. Wiley Interdisciplinary Reviews: Climate Change. 2017. DOI: 10.1002/wcc.535

[8] Roosevelt Island Tidal Energy (RITE) Project Demonstration. Available from: https://tethys.pnnl.gov/annex-iv-sites/roosevelt-island-tidal-energy-rite-project-demonstration

[9] Uihlein A, Magagna D. Wave and tidal current energy–A review of the current state of research beyond technology. Renewable and Sustainable Energy Reviews. 2016;**58**:1070-1081

[10] Das K, Behera RN. A survey on machine learning: Concept, algorithms and applications. International Journal of Innovative Research in Computer and Communication Engineering. 2017;**5**(2): 1301-1309

[11] Hsieh W, Tang B. Applying neural network models top prediction and data analysis in meteorology and oceanography. Bulletin of the American Meteorological Society. 1998;**79**(9): 1855-1870

[12] Nowosad A, Neto A.R. and Campos Velho H. Data assimilation in chaotic dynamics using neural networks. In: International Conference on Nonlinear Dynamics, Chaos, Control and Their Applications in Engineering. 2000; 212–221

[13] Hoang HS, De Mey P, Talagrand O, Baraille R. A new reduced-order adaptive filter for state estimation in high dimensional systems. Automatica. 1997;**33**:1475-1498

[14] Haykin S. Adaptive Filter Theory. Upper Saddle River, NJ: Prentice Hall. ISBN 978-0-13-048434-5; 2002

[15] Patrikar A, Provence J. Nonlinear system identification and adaptive control using polynomial networks. Mathematical and Computer Modelling. 1996;**23**(112):159-173

[16] Haykin S. Neural Networks and Learning Machines. 3rd ed. Upper Saddle River, NJ: Pearson Education; 2009

[17] Cybenko G. Approximations by superpositions of sigmoidal functions. Mathematics of Control, Signals, and Systems. 1989;**2**(4):303-314. DOI: 10.1007/BF02551274

[18] Hoang HS, Talagrandand O, Baraille R. On the design of a stable filter for state estimation in high dimensional systems. Automatica. 2001;**37**:341-359

[19] Hoang HS, Baraille R. On the efficient low cost procedure for estimation of high-dimensional prediction error covariance matrices. Automatica. 2017;**83**:317-330

[20] Spall JC. Introduction to Stochastic Search and Optimization. New York: Wiley; 2003

[21] Hoang HS, Baraille R. Stochastic simultaneous perturbation as powerful method for state and parameter estimation in high dimensional systems. In: Baswell AR, editor. Advances in Mathematics Research. Vol. 20. NY: Nova Science Publishers; 2015. pp. 117-148

[22] Lorenz EN. Deterministic non-periodic flow. Journal of the Atmospheric Sciences. 1963;**20**:130-141

[23] Hoang HS, Baraille R. Prediction error sampling procedure based on dominant Schur decomposition. Application to state estimation in high dimensional oceanic model. Journal of Applied Mathematics and Computing. 2011;**12**:3689-3709

[24] Data.shom.fr. Oceanographic Costal Forecasts. Available from: http://www.shom.fr/en/activities/projects/oceanographic-forecasts/datashom-coastal-oceanographic-forecasts-ibi/

Section 2

Modeling of Renewable Energy Systems

Chapter 4

Static and Dynamic Photovoltaic Cell/Module Parameters Identification

Sid-Ali Blaifi and Bilal Taghezouit

Abstract

The accurate parameters extraction is an important step to obtain a robust PV outputs forecasting for static or dynamic modes. For these aims, several approaches have been proposed for photovoltaic (PV) cell modeling including electrical circuit-based model, empirical models, and non-parametrical models. Moreover, numerous parameter extraction methods have been introduced in the literature depending on the proposed model and the operating mode. These methods can be classified into two main approaches including automatic numerical and analytical approaches. These approaches are commonly applied in the static mode, whereas they can be employed for dynamic parameters extraction. In this chapter, as a first stage, the static parameters extraction for both single and double diodes models is exposed wherein Genetic Algorithm and outdoor measurements are considered for fixed irradiation and temperature. In the second stage, a dynamic parameters extraction is carried out using Levenberg-Marquardt algorithm, where 1 day profile outdoor measurement is considered. After that, the robustness of the proposed approaches is evaluated and the parameters obtained by the static method and that given by the dynamic technique are compared. The test is carried out using 3 days with different weather conditions profiles. The obtained results show that the parameters extraction by dynamic techniques gives satisfactory performances in terms of agreement with the real data.

Keywords: photovoltaic module, static parameters extraction, dynamic parameters extraction, empirical model, electrical model

1. Introduction

The increasing development of PV technologies brought out their potential to provide the energy abundance across the world. Hence, they have been interested by several research groups in the purpose to improve their behavior and extend their life-time. Meanwhile, giving an accurate forecasting of the PV outputs behavior has been always a real issue related to their nonlinearity. Two modes can characterize the PV module in terms of modeling, the first one is the static mode wherein the obtained model is characterized and validated for fixed weather conditions (irradiation and temperature), while the second is the dynamic mode where the validation is carried out using variable weather conditions. In this context, several models of the PV cell/module have been introduced in the literature based mainly on several approaches including electrical, empirical, and non-parametrical modeling. For the non-parametrical models, two approaches are introduced namely, Artificial Neural Network (ANN) and Neuron-Fuzzy based models. The accuracy and the robustness of these approaches rely strongly on the richness of the training dataset in terms of scenarios. Neuron-Fuzzy techniques has been used to predict cell short-circuit current and open-circuit voltage for the static representation [1]. The ANN-based technique is introduced to model the PV array power for the embedded systems implementation [2]. This technique has been tested for dynamic mode dealing [3]. The ANN technique is also used for the prediction of the PV cell/module voltage directed to amorphous silicon PV technology wherein the obtained network has been tested by real dynamic data [4].

In the empirical or the analytical approaches, several models have been proposed to estimate accurately both static and dynamic modes of PV cell/modules. An analytical model is proposed by [5] based on manufacturer characteristic. This model provides acceptable results for both static and dynamic working. Another model has been introduced by Sandia National Laboratory [6], which is widely employed for PV cell/model forecasting especially for the large-scale arrays. Other empirical models have been proposed in order to estimate the PV array power under uniform shading [7, 8].

In the electrical approaches, two models widely prevail owing to their simplicity based on equivalent circuits namely: single (SDM) and double (DDM) diodes-based models. These models can deal with both static and dynamic modes with an acceptable accuracy. Besides, several enhancements have been introduced in these models to minimize parameters number and give more simplicity [9, 10]. The expressions of both photo-generation and diode saturation currents have been improved to give more accuracy in the dynamic working [11].

All models possess unknown parameters, which should be identified according to the module used in the practice. For this end, numerous approaches have been introduced in the literature including analytical and automatic numerical methods. The analytical methods usually rely on specific points on the I-V curve and on some value given by manufacturer. However, a significant error can be engendered if one of more of selected points is incorrect [12].

Owing to their adequate results, automatic numerical methods are prevail in PV models parameters extraction either through the use of deterministic algorithms such as: Newton model modified with Levenberg [13], Levenberg-Marquardt [14], Simulated Annealing algorithm (SA) [15], Pattern Search (PS) [16], Nelder-Mead Simplex algorithm (NMS) [17], and hybrid Nelder-Mead and modified particle swarm optimization [18] or by introducing metaheuristic algorithms such as: Genetic Algorithms (GA) [19], Particle Swarm Optimization (PSO) [20–26], Cuckoo Search (CS) [27], Artificial Bee Colony (ABC) [26, 28], and Artificial Bee Swarm (ABS) [29]. Moreover, other algorithms have been introduced like (FPA) [30, 31], hybrid Bee Pollinator Flower Pollination Algorithm (BPFPA) [31, 32], Harmony Search (HS) [33], Artificial Fish Swarm Algorithm (AFSA) [34], and other algorithms. The majority of the aforementioned algorithms have been applied for static parameters extraction.

Numerical algorithms have been also applied for dynamic parameters extraction wherein the identification process is carried out using variables weather conditions [35, 36].

In this chapter, modeling and parameters extraction of PV cell/module are detailed. Where, comparison study among three models by applying static and dynamic identification using out-door measurement.

2. PV cell/module modeling

The PV cell presents outputs variation, which depends on weather conditions namely, irradiation and temperature. As illustrated in **Figure 1**, for load variation from open circuit to short circuit, the PV cell shows nonlinear characteristic that possess a maximum point of power. For an optimal working, the load should be adapted at this point. In this section, three PV cell models will be employed and improved using automatic parameters extraction namely: the empirical Sandia model and both single and double diodes electrical models.

2.1 Sandia model

This empirical model given by SANDIA National Laboratories provides relatively accurate dynamic forecast for PV cell/module by describing the thermal, the electrical, and the optical characteristics. Also, this model can be destined for any technology and can be adapted with any scale of PV arrays. Furthermore, its simplicity can qualify it to be used for real-time online prediction. Expressions (1)–(4) describe the variation of I_{mpp}, V_{mpp}, and P_{mpp}, respectively.

$$I_{mp} = I_{mp\_STC}\left(C_0E_e + C_1E_e^2\right)\left(1 + \alpha_{mp}(T - T_{STC})\right) \tag{1}$$

$$V_{mp} = V_{mp\_STC} + C_2N_s \times \delta(T) \times \ln(E_e) + C_3N_s(\delta(T)\ln(E_e))^2 + \beta_{mp}(T - T_{STC}) \tag{2}$$

$$\delta(T) = \frac{n \times k \times (T + 273.15)}{q} \tag{3}$$

$$P_{mp} = I_{mp} \times V_{mp} \tag{4}$$

where, C_{0-3} are empirical parameters to be identified, I_{mp_STC}, V_{mp_STC} are the current and the voltage in the maximum power point under standard test condition, E_e is the effective irradiation, K is the Boltzmann constant, q is the electron charge, δ(T) is the thermal voltage, α_{mp} and β_{mp} are, respectively, the current and the voltage temperature coefficient [36].

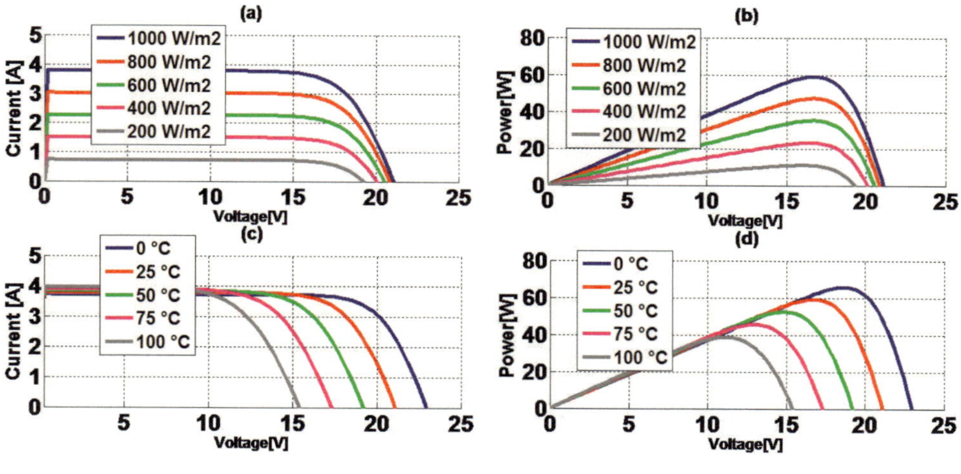

Figure 1.
PV characteristic for different irradiation and temperature. (a) Current versus voltage for different irradiation; (b) Power versus voltage for different irradiation; (c) Current versus voltage for different temperature; (d) Power versus voltage for different temperature.

2.2 Single diode based model (SDM)

This physical model is based on the electrical approach illustrated in **Figure 2** wherein the PV cell is composed of: a photo-generation current source and a diode while joule losses are represented by two serial and parallel resistors.

From this electrical representation, expression (5) and (6) can be obtained to describe the evolution of both current and voltage. The output current is expressed as a sum of the photo-generation current I_{ph}, the diode current I_d, and the shunt current I_{sh}.

$$I = I_{ph} - I_d - I_{sh} \tag{5}$$

$$I_d = I_0 \left[\exp\left(\frac{q(V + R_s I)}{nKT} \right) - 1 \right] \tag{6}$$

$$I_{sh} = \frac{V + R_s I}{R_{sh}} \tag{7}$$

where, K is the constant of Boltzmann, q is the electron charge, T is the cell temperature, n is the diode ideality factor, and I_0 is the current saturation due to diffusion and recombination.

After the substitution of Eqs. (6) and (7) in (5), the following expression is obtained:

$$I = I_{ph} - I_0 \left[\exp\left(\frac{q(V + R_s I)}{nKT} \right) - 1 \right] - \frac{V + R_s I}{R_{sh}} \tag{8}$$

R_s, R_{sh}, and n are parameters to be identified in the static study and can be adjusted in the dynamic study.

Diode saturation current I_0 is expressed in Eq. (9) function of the cell temperature and the energy band-gap [11].

$$I_0 = \frac{I_{sc} \times T^3 \times \exp\left(-\frac{q \times E_g}{KT}\right)}{\left(\exp\left(\frac{q \times V_{oc}}{nKT_r}\right) - 1\right) \times T_r{}^3 \times \exp\left(-\frac{q.E_{gr}}{KT_r}\right)} \tag{9}$$

Eq. (10) describes the evolution of the energy band-gap E_g as function of the cell temperature.

$$E_g = E_{g0} - \left(\frac{\alpha \times T^2}{\beta + T} \right) \tag{10}$$

where E_{g0} and E_{gr} are the energy band-gap of the silicon at 0°C and at the reference temperature T_r, respectively, α and β are constants of the material.

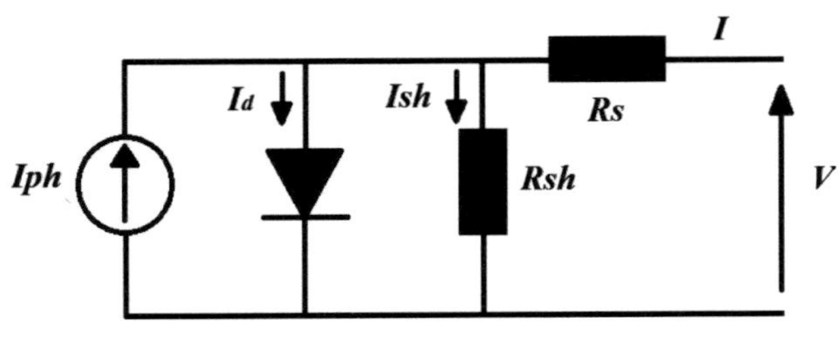

Figure 2.
Single diode equivalent circuit.

The photo-generation current is given by Eq. (11) as a function of the irradiation and the cell temperature.

$$I_{ph} = \frac{G}{1000}\left[I_{sc} + \mu(T - T_r)\right] \tag{11}$$

where G is the input irradiation, T is the cell temperature, I_{sc} is the module short-circuit current, and μ is the coefficient temperature/short-circuit current (given by the manufacturer) [11].

E_{g0}, α, β, μ, I_{sc}, and V_{oc} are parameters to be identified in the dynamic study.

2.3 Double diode-based model

From the electrical representation illustrated in **Figure 3**, the PV cell can be represented by a source of current that represents the photo-generation, two diodes and both parallel and serial resistances representing the loss of energy inside the cell.

After applying nodes law, the output current is expressed as sum of: photo-generation current I_{ph}, shunt current I_{sh} and the diodes currents I_{d1} and I_{d2} (Eqs. (12)–(15)).

$$I = I_{ph} - I_{d1} - I_{d2} - I_{sh} \tag{12}$$

$$I_{d1} = I_{01}\left[\exp\left(\frac{q(V + R_s I)}{n_1 KT}\right) - 1\right] \tag{13}$$

$$I_{d2} = I_{02}\left[\exp\left(\frac{q(V + R_s I)}{n_2 KT}\right) - 1\right] \tag{14}$$

$$I_{sh} = \frac{V + R_s I}{R_{sh}} \tag{15}$$

In which I_{01-2} are currents saturation of the two diodes that resulted from diffusion and recombination, n_{1-2} are ideally factors.

By substituting Eqs. (13)–(15) in (12), final description of the output current versus the voltage is obtained which is expressed in Eq. (16) [11].

$$I = I_{ph} - I_{01}\left[\exp\left(\frac{q(V + R_s I)}{n_1 KT}\right) - 1\right] - I_{02}\left[\exp\left(\frac{q(V + R_s I)}{n_2 KT}\right) - 1\right] - \frac{V + R_s I}{R_{sh}} \tag{16}$$

n_{1-2} and R_s, R_{sh} are parameters which will be identified in the static study and they can be adjusted in the dynamic study.

Eqs. (17) and (18) express the evolution saturation currents of the diodes I_{01-2} versus energy band-gap E_g and cell temperature [11].

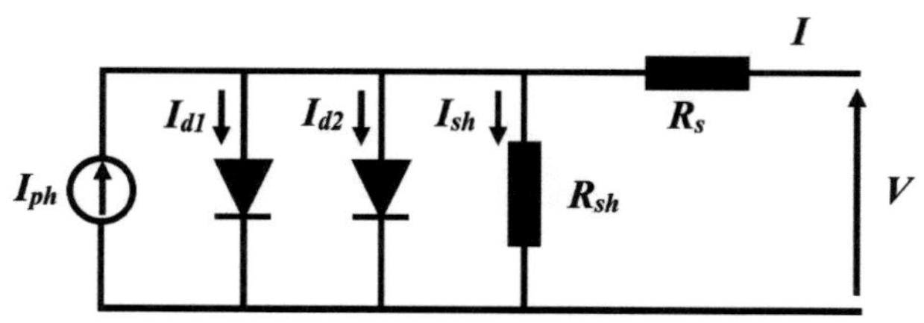

Figure 3.
Double diodes equivalent circuit.

$$I_{01} = \frac{I_{sc} \times T^3 \times \exp\left(-\frac{q \times E_{g1}}{KT}\right)}{\left(\exp\left(\frac{q \times V_{oc}}{n_1 K T_r}\right) - 1\right) \times T_r^{\ 3} \times \exp\left(-\frac{q \times E_{gr}}{K T_r}\right)} \tag{17}$$

$$I_{02} = \frac{I_{sc} \times T^3 \times \exp\left(-\frac{q \times E_{g2}}{KT}\right)}{\left(\exp\left(\frac{q \times V_{oc}}{n_2 K T_r}\right) - 1\right) \times T_r^{\ 3} \times \exp\left(-\frac{q \times E_{gr}}{K T_r}\right)} \tag{18}$$

The photo-generation current is represented by the same expression of the single diode model (Eq. (11)).

Parameters E_{g01-2}, α_{1-2}, β_{1-2}, μ, I_{sc} and V_{oc} will be identified in the dynamic study.

3. Static parameters extraction of PV module

For fixed irradiation and temperature, a static parameters extraction will be done to extract five parameters in SDM and seven parameters in DDM. A numerical stochastic optimization algorithm is used in this identification. This algorithm namely, Genetic Algorithm (GA), is employed to minimize the cost function given in Eq. (19) which expresses the root mean square error (RMSE) between the measured PV module I(v) characteristic and that given by the models. For this and, outdoor static measurements have been carried out using the peak measuring device tracer (PVPM 2540C), whose characteristics are illustrated in **Table 1**. This device has been programed to provide both I(V) and P(V) curves of 101 samples per 1 min.

$$RMSE = \sqrt{\frac{1}{N}\sum\nolimits_{i=1}^{N} f(V_m, I, P)^2} \tag{19}$$

where I is the simulated current, V_m is the measured voltage, N is the number of sample in I(V) characteristics. The error between the measured and simulated I(V) characteristics for the aforementioned models are expressed in Eqs. (20) and (21).

$$f(V_m, I, P) = I_{ph} - I_0\left[\exp\left(\frac{q(V_m + R_s I)}{nKT}\right) - 1\right] - \frac{V_m + R_s I}{R_{sh}} - I_m \tag{20}$$

$$f(V_m, I, P) = I_{ph} - I_{01}\left[\exp\left(\frac{q(V_m + R_s I)}{n_1 KT}\right) - 1\right] - I_{02}\left[\exp\left(\frac{q(V_m + R_s I)}{n_2 KT}\right) - 1\right] - \frac{V_m + R_s I}{R_{sh}} - I_m \tag{21}$$

Application	DC voltage	DC current	Temperature	Irradiance	Measuring period single measurement	I-V curve samples
PV modules and small strings	25/50/ 100/250 V	2/5/10/ 40 A	−40°C to +120°C with Pt1000	0–1300 (W/m^2) (standard-sensor)	0.02–2 (s)	101

Table 1.
PVPM2540C characteristics.

3.1 Genetic Algorithm

The Genetic Algorithm (GA) is a stochastic algorithm imitated from the biological genetic process used to find an approximate solution for optimization problems. Like in the natural concept, the chromosome is the holder of the genes that the child can probably get from his parents. By analogy, these genes represent the variables (parameters) of the function to be minimized. Five steps can characterize the GA namely, generation of initial population, evaluation of fitness, selection, crossover and mutation [37, 38].

3.1.1 Initial population

The process starts by the generation the initial population of N chromosome coded in binary. Each vector chromosome is formed of group of parameters in which its length M is given in Eq. (22) wherein n is the number of parameters and N_b is the length of the sub-string (number of bits) of each parameter as shown in **Figure 4**. The length of the integer part given by the vector Conv (Eq. (23)) is used to limit the research domain in which, P_i (Eq. (24)) is the parameter value in decimal code [37, 38].

$$M = n \times N_b \tag{22}$$

$$Conv = \left[2^{n_i}\, 2^{n_i-1} \ldots 2^0\, 2^{-1} \ldots 2^{(n_i - N_b + 1)}\right] \tag{23}$$

$$P_i = \left[a_0\, a_1 \ldots a_{N_b-1}\right] \times Conv^T \tag{24}$$

3.1.2 Fitness

In this stage, the parameters values that have been randomly generated and decoded in decimal base will be substituted in the cost function to be optimized. The fitness is the solution of the parameters in the RMSE (x) function calculated in Eq. (19). Its value is mathematically expressed in Eq. (23) [37, 38].

$$Fitness(x) = \frac{1}{1 + RMSE(x)} \tag{25}$$

3.1.3 Selection

The chromosomes that will participate as parents to generate a new child are chosen in this step. Any chromosome in the generated population can be chosen however, the individual that presents a good fitness have a high probability. The technique used for the chromosome choice is the roulette wheel illustrated in **Figure 5**, wherein the selection probability P_s, expressed in Eq. (24), is calculated, consists of a cumulative sum of the fitness of each chromosome orderly relative to the sum of all fitness. After that, the process generates a random drawing probability Pr. Hence, the first chromosome corresponds to $P_r < P_s$ is chosen for the next steps (crossover and Mutation).

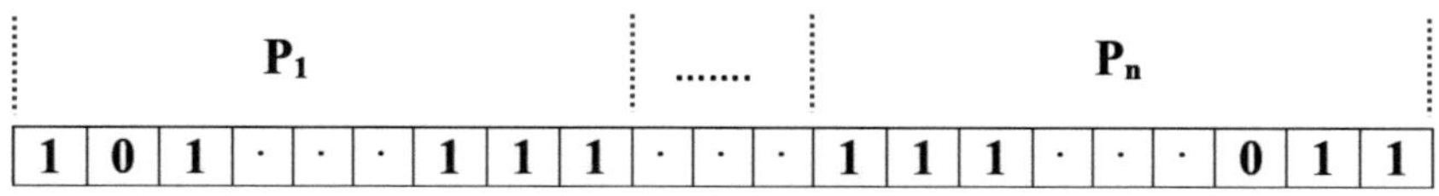

Figure 4.
Chromosome string.

$$P_s(i) = \frac{\sum_{k=1}^{i} Fitness(k)}{\sum_{j=1}^{N} Fitness(j)} \tag{26}$$

where, k is an integer counter that varies from 1 to the current chromosome, and j is an integer counter that varies from 1 to the population size N [37, 38].

3.1.4 Crossover

After selecting the chromosomes, the algorithm gives birth to new children by performing a crossover between each two chromosomes. For this end, a drawing probability P_r is generated and compared with the crossover probability P_c (usually high probability). Hence, the parents chromosomes that corresponds to $P_r < P_c$ will be chosen for child generation. If not, the same chromosomes are kept. As illustrated in **Figure 6**, the crossover by point is used wherein the bits after the point randomly chosen are swapped [37, 38].

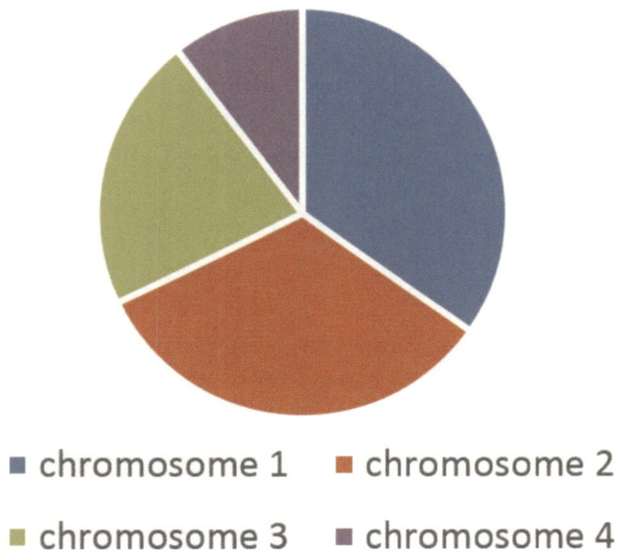

Figure 5.
Roulette wheel example.

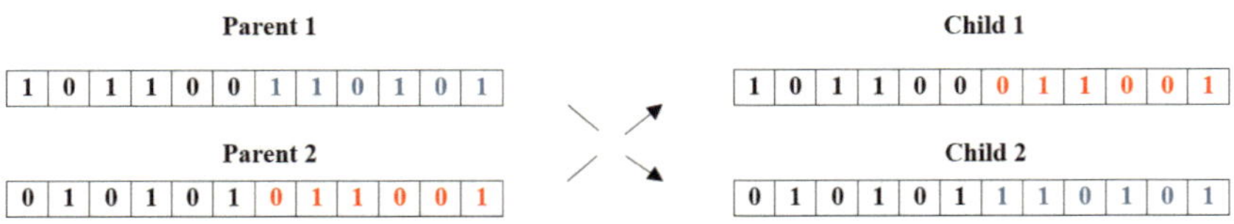

Figure 6.
Crossover process.

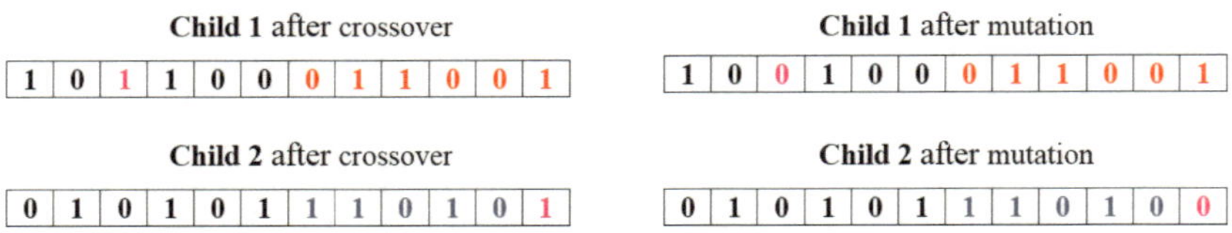

Figure 7.
Mutation process.

3.1.5 Mutation

In this step, the algorithm introduces a change in some characters of the selected chromosomes in order to expand the search space if the initial population does not

Description	SANYO mono-crystalline
Cell number	96
Cell type	Mono-crystalline
Cell size	156 × 156 mm
PV module dimension	1319 × 894 × 35 mm
Nominal power	180 W
Open circuit voltage V_{oc}	66.4 V
Short circuit current I_{sc}	3.65 A
Voltage V_{mpp}	54 V
Current I_{mpp}	3.33 A
Nominal operating temperature NOCT	45 ± 2°C
Temperature coefficient (P_{max})	−0.33%/°C
Temperature coefficient (I_{sc})	1.10 mA/°C
Temperature coefficient (V_{oc})	−0.173 V/°C

Table 2.
PV modules characteristics.

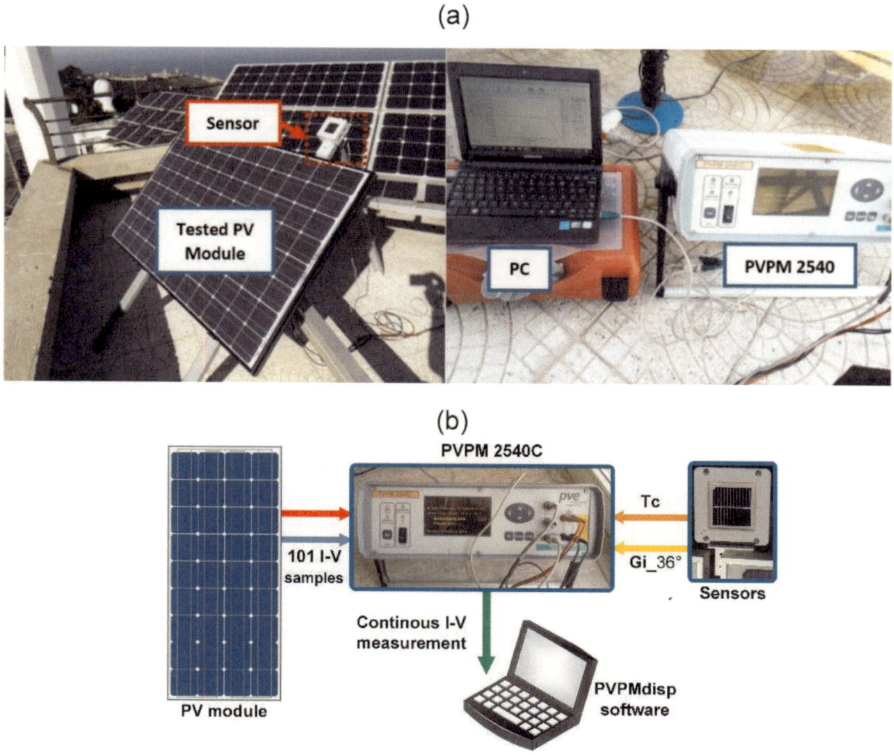

Figure 8.
(a) Pictures for I-V experimental platform and (b) synoptic scheme for I-V measurement.

fall in the optimal solution. In the binary coding, the selected bit will change from 1 to 0 and vice versa as described in **Figure 7**. The mutation has low probability P_m in which, it will be affected for characters that correspond to $P_r < P_m$ in which, P_r is the drawing probability (randomly generated) [37, 38].

Our system is formed of mono-crystalline PV module SANYO technology with the characteristics listed in **Table 2**, peak measuring tracer and the necessary sensors. The experimental platform is illustrated in both **Figure 8a** and **b**.

After running of GA for 1000 cycles with the parameters listed in **Table 3** for both SDM and DDM using outdoor measurement of the systems, wherein extracted parameters are listed in **Table 4**. **Figure 9a** and **b** illustrate the agreement between the measured and simulated I(V) and P(V) characteristics for SDM model whose

GA parameters	Value
Number of cycle	1000
Population length	500
Crossover probability	0.7
Mutation probability	0.2

Table 3.
The GA parameters.

The electrical parameter	I_{ph} [A]	I_0 [A]	n	Rs [Ω]	Rsh [KΩ]
The identified value	3.0195	49591e−005	1.874*96	0.3273	8.1514

Table 4.
The obtained parameters for SDM.

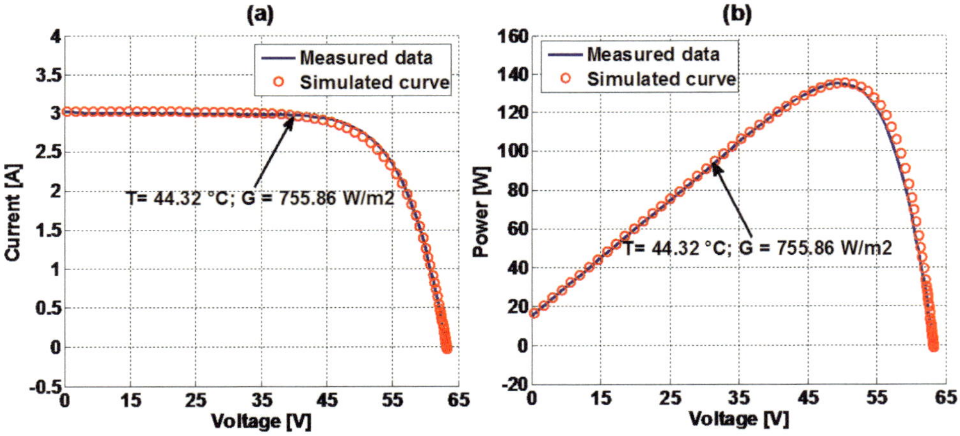

Figure 9.
Matching between measured and simulated characteristics for SDM. (a) I(V) and (b) P(V).

The electrical parameter	I_{ph} [A]	I_{01} [A]	I_{02} [A]	n_1	n_2	Rs [Ω]	Rsh [KΩ]
The identified value	3.0289	6.1035e−005	3.8147–006	1.3658*96	1.9179*96	0.1017	5.992

Table 5.
The obtained parameters for DDM.

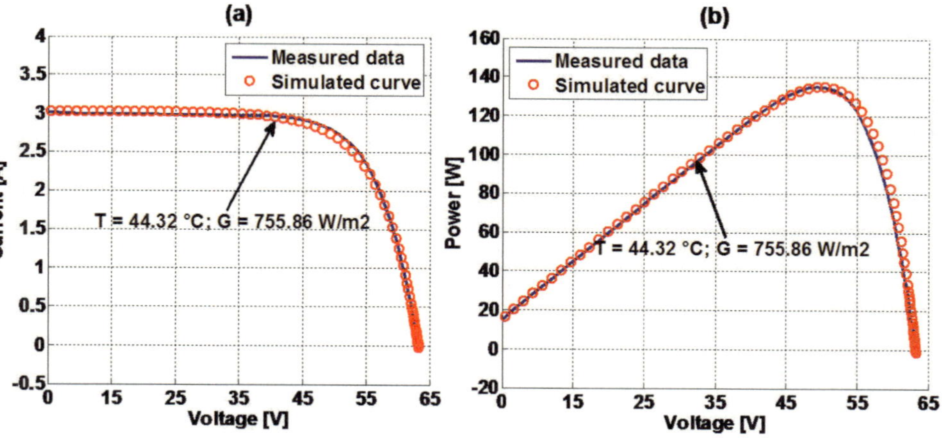

Figure 10.
Matching between measured and simulated characteristics for DDM. (a) I(V) and (b) P(V).

E	*ΔE*				
	NS	*NL*	*ZE*	*PL*	*PS*
NS	*PS*	*PS*	*PL*	*PS*	*PS*
NL	*PS*	*PL*	*PL*	*PL*	*PS*
ZE	*NL*	*NL*	*ZE*	*PL*	*PL*
PL	*NS*	*NL*	*NL*	*NL*	*NS*
PS	*NS*	*NS*	*NL*	*NS*	*NS*

Table 6.
The inference table.

obtained parameters are summarized in **Table 5**. **Figure 10a** and **b** show the agreement between the measured and simulated I(V) and P(V) characteristics for DDM model whose extracted parameters are summarized in **Table 6**.

Some parameters will be identified again in the dynamic study including the parameters involved in I_0 and I_{ph} equations, while the remaining will be adjusted to give more accuracy under variable weather conditions.

4. Dynamic parameters extraction of PV module

In this section, dynamic parameters identification will be described wherein the process is done by using 1 day profile of measurement. This allows to improve the nominal values given by the manufacturer, which can cause a significant error due to operating conditions and the consumed lifetime. Moreover, parameters obtained by static method can be adjusted by dynamic identification. For this end, automatic parameters adjustment using Levenberg-Marquardt optimization algorithm is employed.

As illustrated in **Figure 11**, the main idea is to take both PV module model and the MPPT as a single system with three outputs namely, I_{mpp}, V_{mpp}, and P_{mpp}. These outputs will be compared with 1 day profile of outdoor measurements. The process consists in minimizing the error between the model outputs and the real data. The whole system has been implemented in Matlab/Simulink tool.

For Sandia model, the process is carried out without the use of MPPT considering that this model has been established to the dynamic forecasting.

4.1 The MPPT used

An Accurate fuzzy logic MPPT algorithm is employed in our system (for SDM and DDM) in order to get satisfactory results in terms of precision and accuracy.

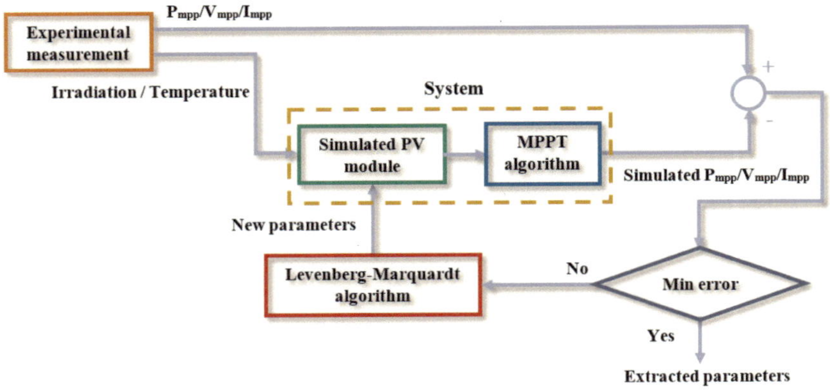

Figure 11.
Dynamic parameters extraction process.

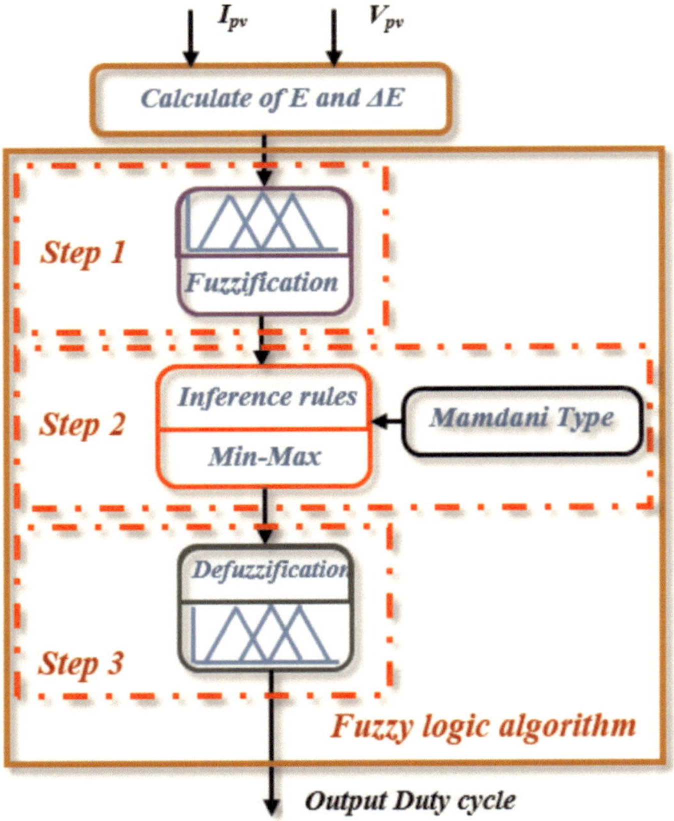

Figure 12.
Fuzzy logic algorithm steps.

The algorithm is used to control a DC/DC boost converter for the purpose to keep the PV module working at the maximum point of power. Mamdani inference model is used with two inputs namely, the error E and the variation of the error ΔE. The calculation of these attributes is expressed in Eqs. (27)–(30).

$$\Delta P_{pv}(n) = P_{pv}(n) - P_{pv}(n-1) \tag{27}$$

$$\Delta V_{pv}(n) = V_{pv}(n) - V_{pv}(n-1) \tag{28}$$

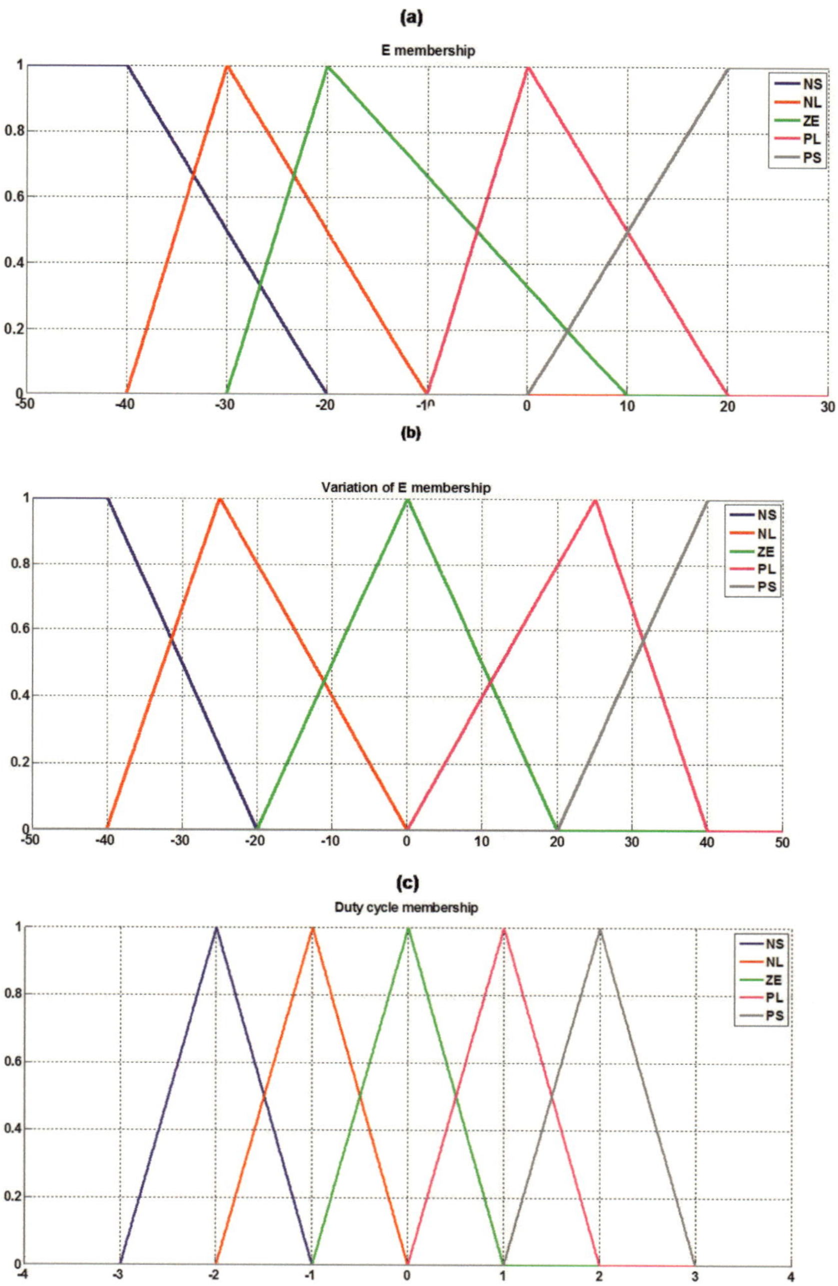

Figure 13.
(a) Error membership; (b) variation of error membership; and (c) duty cycle membership.

$$E(n) = \frac{\Delta P_{pv}(n)}{\Delta V_{pv}(n)} \tag{29}$$

$$\Delta E(n) = E(n) - E(n-1) \tag{30}$$

Three steps can characterize the fuzzy algorithm; the first one is the fuzzification process that consists on the conversion of the numerical inputs values (E and ΔE) into linguistic values by the substitution in the membership functions. The second step is the inference process, which is considered as the main stage in the fuzzy algorithm wherein the relation between the inputs and the output is done. The third step is the defuzzification where the process converts the linguistic decision into numerical output. **Figure 12** describes briefly the fuzzy processing steps [39].

SDM parameters	Boundaries	DDM parameters	Boundaries	Sandia parameters	Boundaries
n	[0, 2]*96	n_1, n_2	[0, 2]*96	C_0	[0, 2]
R_s [Ω]	[0, 1]	R_s [Ω]	[0, 1]	C_1	[−1, 1]
R_{sh} [Ω]	[0, 10^4]	R_{sh} [Ω]	[0, 10^4]	C_2	[−10, 10]
E_{g0}	[0, 1]	E_{g01-2}	[0, 2]	C_3	[−10, 50]
A	[0, 1]	α_{1-2}	[0, 1]	α_{Imp} [°C^{-1}]	[0, 1]
B	[0, 10^4]	β_{1-2}	[0, 10^4]	β_{Vmp} [V/°C]	[−1, 0]
μ	[0, 1]	μ	[0, 1]	n	
I_{sc}	[3, 3.7]	I_{sc}	[3, 3.7]		
V_{oc}	[60, 66.8]	V_{oc}	[60, 66.8]		

Table 7.
Lower and upper boundaries selected for each model parameters.

SDM parameters	Values	DDM parameters	Values	Sandia parameters	Values
n	105.73/96	n_1, n_2	90.73/96; 73.39/96	C_0	1.058
R_s [Ω]	0.82495	R_s [Ω]	0.3219	C_1	0.020
R_{sh} [Ω]	8.371×10^3	R_{sh} [Ω]	4.9664×10^3	C_2	−0.341
E_{g0} [ev]	1.4525	E_{g01-2} [ev]	1.649; 1.31	C_3	−9.997
α	6.56×10^{-4}	α_{1-2}	0.0018; 0.0132	α_{Imp} [°C^{-1}]	2.53×10^{-14}
β	126.11	β_{1-2}	694.84; 1020.76	β_{Vmp} [V/°C]	−0.203
μ	0.0121	μ	0.0112	n	1.221
I_{sc}	3.671	I_{sc}	3.629		
V_{oc}	66.208	V_{oc}	65.527		

Table 8.
Dynamic extracted parameters.

For both inputs and output, five trapezoidal and triangular membership functions have been employed namely: NS (negative strong), NL (negative low), ZE (zero), PL (positive low) and PS (positive strong). The center of gravity based method is used for the defuzzification to provide the control duty cycle after applying the Mamdani inference model given in **Table 6**. **Figure 13a, b** and **d** describes the used membership functions [39].

4.2 Simulation study

The Levenberg-Marquardt algorithm is implemented using 1 day profile of outdoor real measurement of dynamic PV outputs (P_{mpp}, I_{mpp} and V_{mpp}). The process consists in minimizing the error between simulated outputs of both SDM and DDM and 8 h of real data (09:00 am–05:00 pm). The peak measuring device tracer (PVPM 2540C) has been programmed to provide 1 sample per minute. **Table 7** lists the lower and upper limits search of the extracted parameters. The extracted parameters using the dynamic method are summarized in **Table 8**. The inputs measurement of the irradiation and the temperature are illustrated in **Figure 14a** and **b**, respectively. Satisfactory results have been obtained in terms of matching between the real data and the simulated outputs P_{mpp}, I_{mpp} and V_{mpp} for SDM, DDM and Sandia as shown in **Figure 15a–c**, respectively.

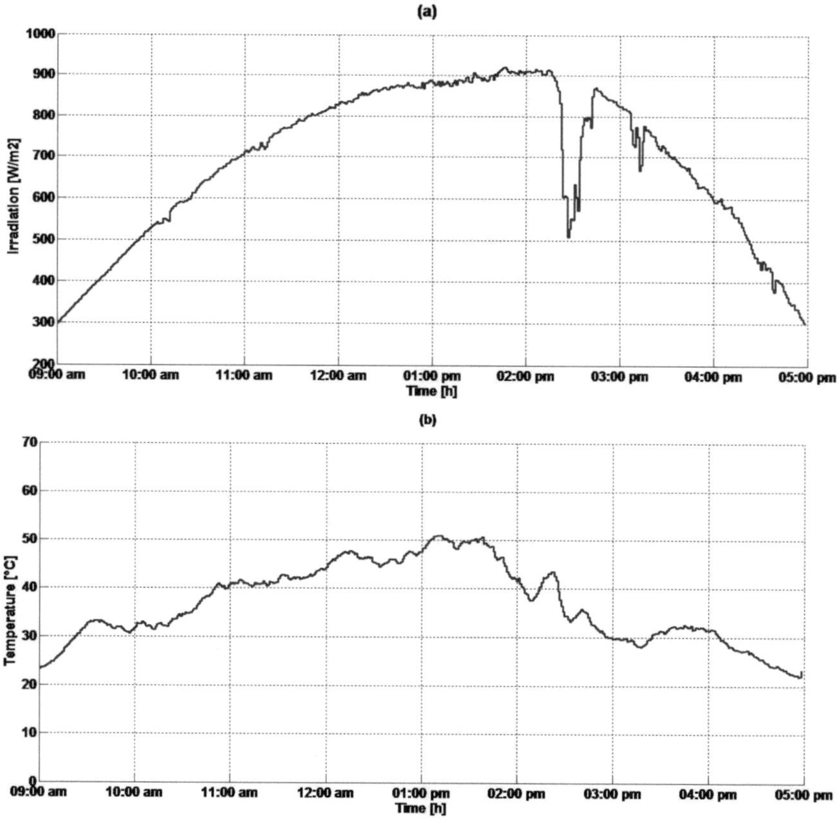

Figure 14.
Eight hours profile (09:00 am–05:00 pm). (a) Irradiation and (b) temperature.

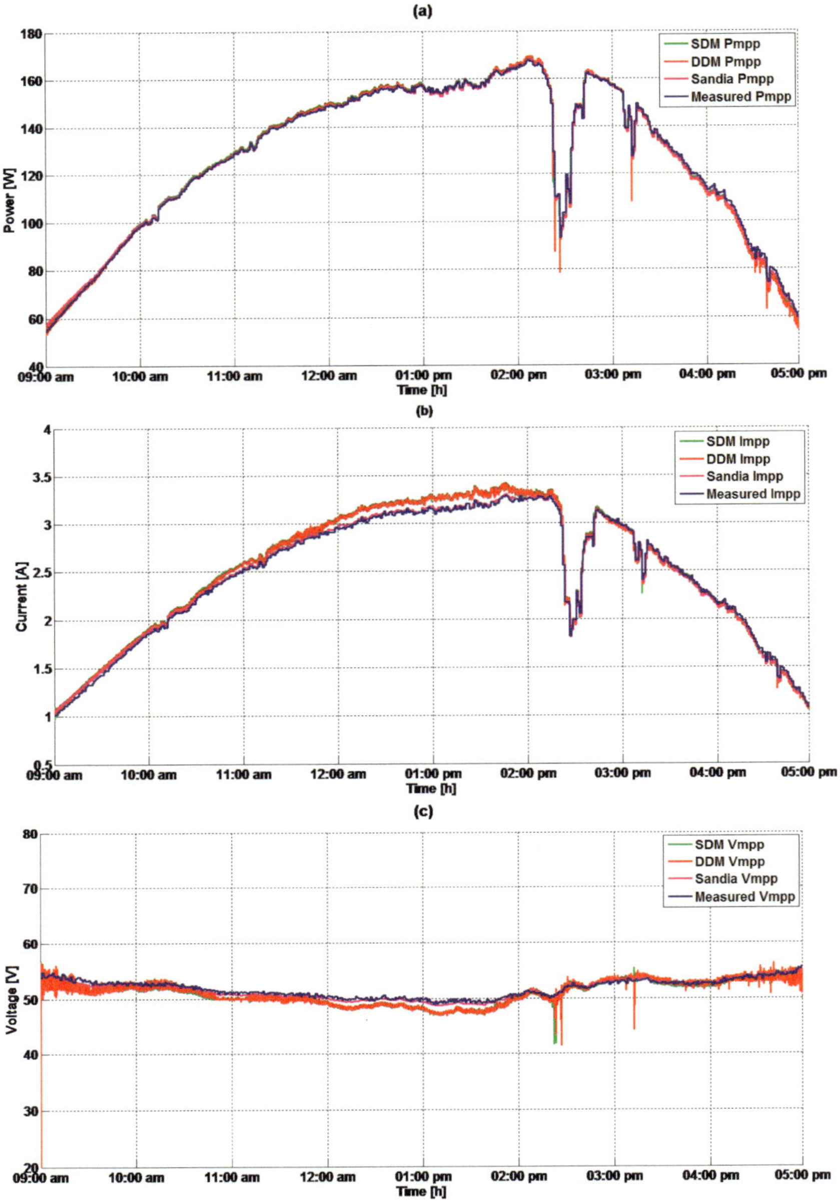

Figure 15.
Eight hours profile (09:00 am–05:00 pm) of: measured output VS obtained output for SANYO mono-crystalline. (a) Measured P_{mpp} vs obtained P_{mpp} from: SDM, DDM and Sandia, (b) measured I_{mpp} vs obtained I_{mpp} from: SDM, DDM and Sandia and (c) measured V_{mpp} VS obtained V_{mpp} from: SDM, DDM and Sandia.

5. Experimental validation

In this section, a validation with an unseen data is carried out to test and compare the effectiveness of the proposed enhancement. The three developed models will be compared with real measurement profile (09:00 am–05:00 pm) of irradiation and temperature for different weather conditions. Wherein, the SDM and DDM models using the developed parameters are compared with the former nominal parameters listed in **Table 9**, Sandia model and the real data. Three different skies of real measurement have been used for this validation namely, clear day,

Parameters	Value
E_{g0}	1.16 eV
α	4.73×10^{-4}
β	1000
μ	1.10 mA/°C
I_{sc}	3.65 A
V_{oc}	66.4 V

Table 9.
Nominal parameters.

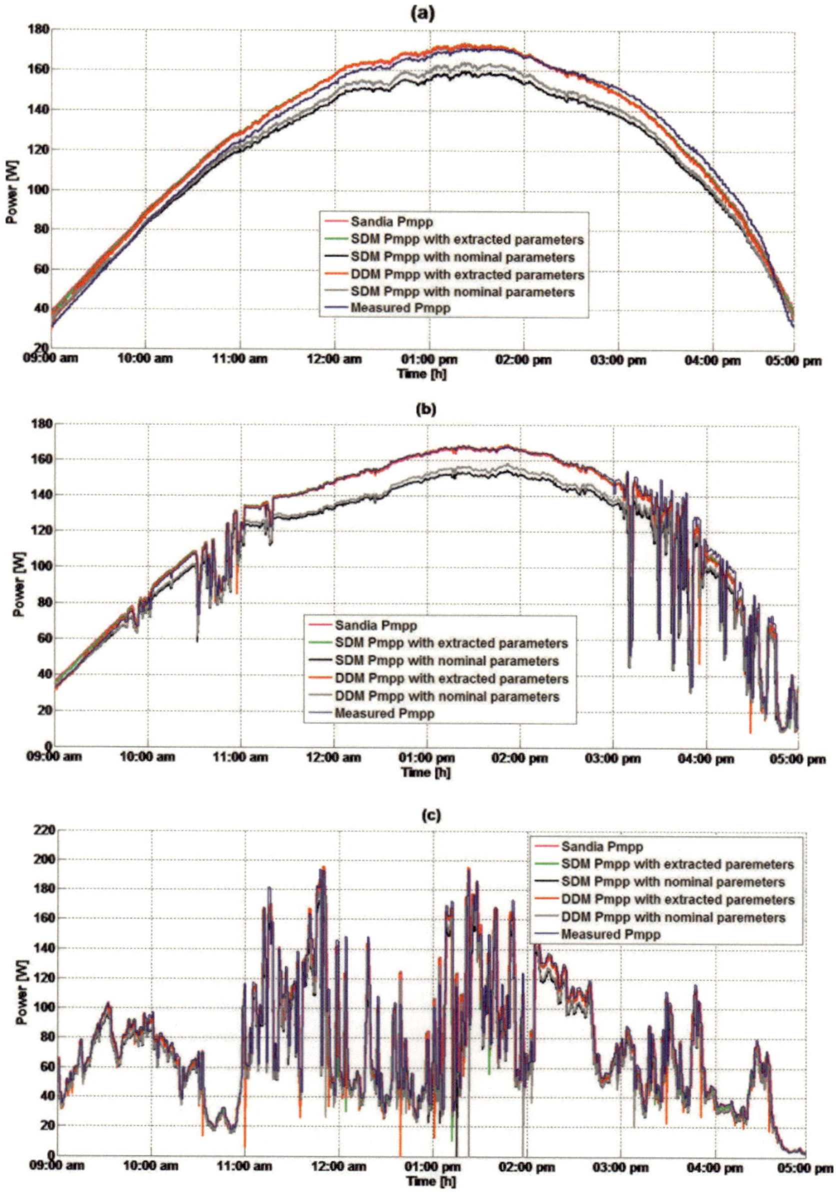

Figure 16.
Eight hours profile (09:00 am–05:00 pm) of: measured P_{mpp} vs obtained P_{mpp} from: SDM, SDM with nominal parameters, DDM, DDM with nominal parameters and Sandia. (a) clear day, (b) semi-cloudy day and (c) cloudy day.

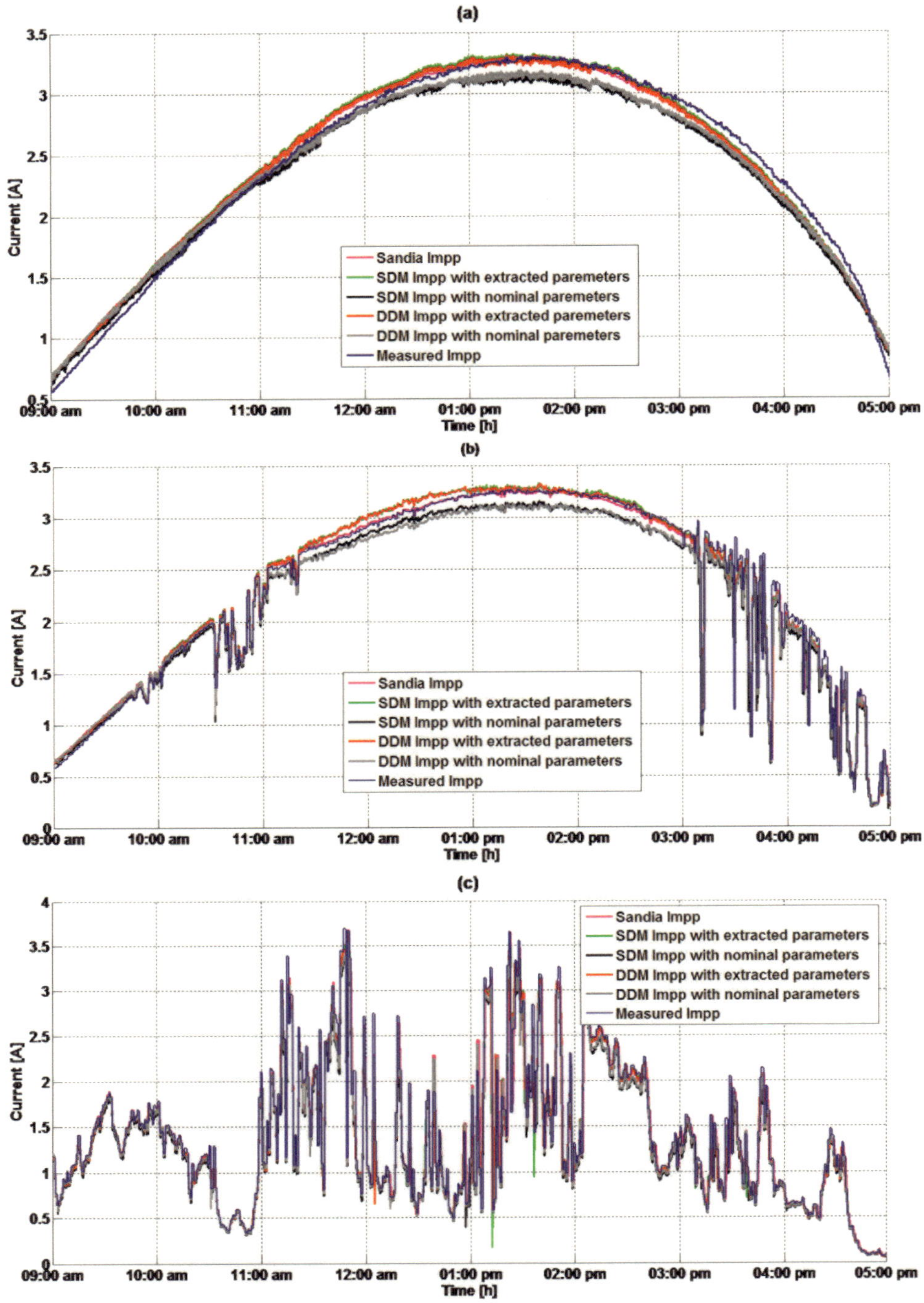

Figure 17.
Eight hours profile (09:00 am–05:00 pm) of: measured I_{mpp} vs obtained I_{mpp} from: SDM, SDM with nominal parameters, DDM, DDM with nominal parameters and Sandia. (a) clear day, (b) semi-cloudy day and (c) cloudy day.

semi-cloudy day and cloudy day. The matching in the power (P_{mpp}) between and the real data and SDM and DDM with nominal parameters, SDM and DDM with the new parameters and Sandia model is illustrated in **Figure 16a–c** for clear day, semi-cloudy day and cloudy day respectively. Besides, the agreement in the voltage and the current (V_{mpp} and I_{mpp}) for these models with the real data is shown in **Figure 17a–c** and **Figure 18a–c**, respectively.

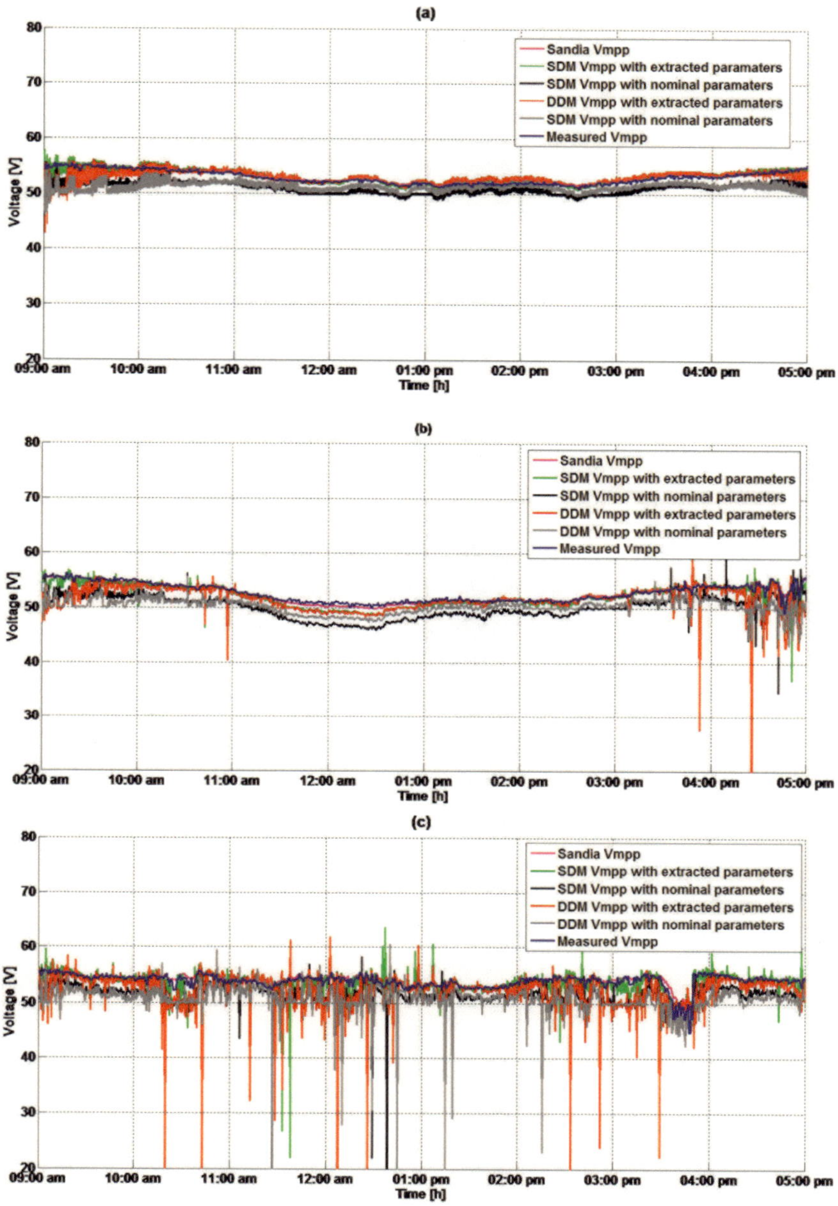

Figure 18.
Eight hours profile (09:00 am–05:00 pm) of: measured V_{mpp} vs obtained V_{mpp} from: SDM, SDM with nominal parameters, DDM, DDM with nominal parameters and Sandia. (a) clear day, (b) semi-cloudy day and (c) cloudy day.

It is clearly found that an improved agreement has been shown by models with new parameters compared to that given by the nominal parameters and the static method.

For more clarity, the hourly power efficiency given by the presented models and the real data has been calculated. It consists on the average of the power during 1 h versus the optimal PV module power (Eq. (31)) [40].

$$\text{Hourly efficiency} = \frac{\sum_{k=1}^{N} P_{MPP}(k)/\text{N}}{P_{PV\_optimal}} \tag{31}$$

in which, N is the number samples per hour.

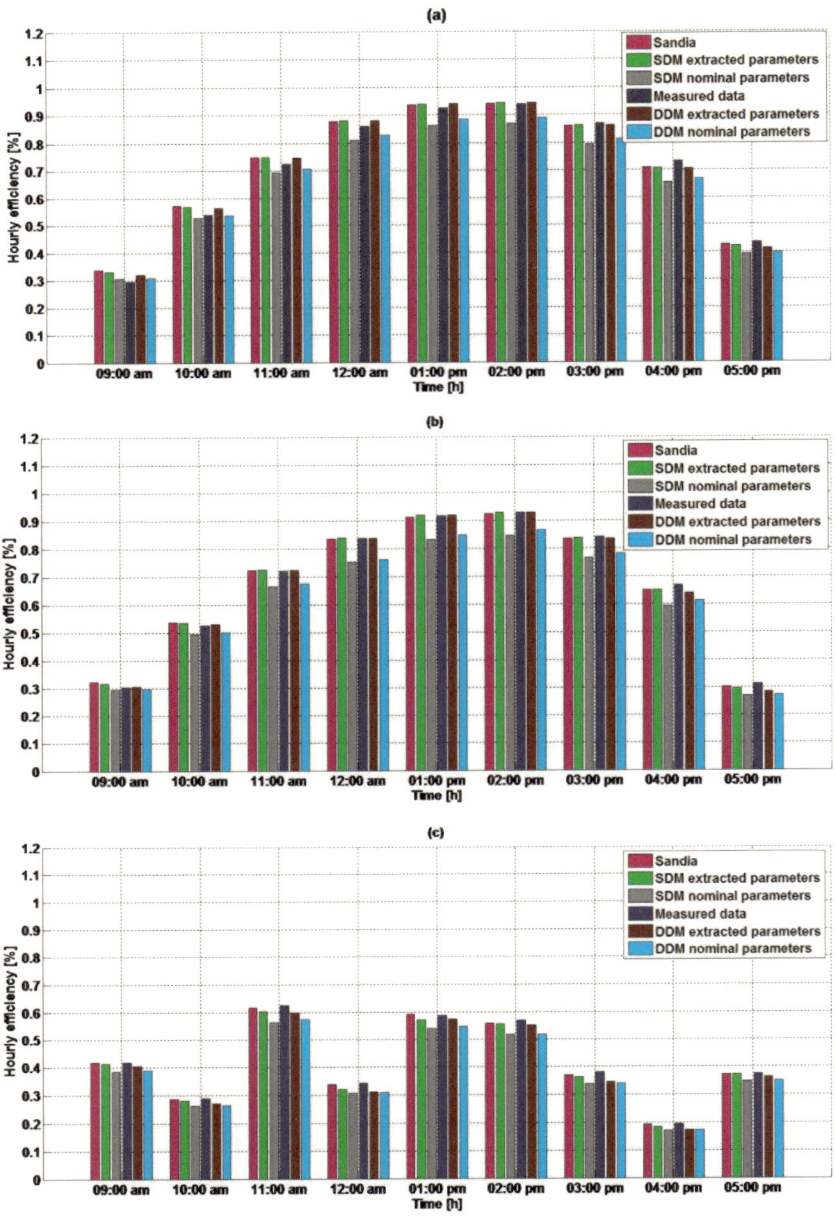

Figure 19.
Eight hours profile (09:00 am–05:00 pm) of: measured hourly efficiency vs obtained hourly efficiency from: SDM, SDM with nominal parameters, DDM, DDM with nominal parameters and Sandia. (a) Clear day, (b) semi-cloudy day, and (c) cloudy day.

Figure 19a–c show the bar-graph of the hourly power efficiency of the proposed models for the three weather conditions, namely the clear day, semi-cloudy day and cloudy day, respectively. The enhanced models present higher hourly power efficiency versus models with the former parameters and those given by the static technique. Furthermore, root mean square error (RMS) and the mean absolute error (MAE) between the real data and the studied models are calculated by Eqs. (19) and (32) to show the enhancement of the proposed method.

$$MAE = \frac{1}{N}\sum_{i=1}^{N}\frac{|Simulated\ \ output - Measured\ \ output|}{Measured\ \ output} \tag{32}$$

Day	Weather	Error [%]	Current					Voltage				
			SDM	SDMnp	DDM	DDMnp	Sandia	SDM	SDMnp	DDM	DDMnp	Sandia
22/01/2018	Clear day	RMS	0.72	1.12	0.69	0.93	0.68	1.48	2.21	1.63	2.34	1.49
		MAE	4.31	5.05	4.24	5.12	4.38	1.27	4.07	1.77	3.70	1.17
29/01/2018	Semi-cloudy	RMS	0.22	0.45	0.24	0.32	0.23	3.30	9.10	6.53	7.87	1.13
		MAE	3.14	5.40	2.96	4.35	2.85	1.50	5.34	2.37	4.76	0.57
28/01/2018	Cloudy day	RMS	0.43	0.48	0.47	0.31	0.52	9.05	8.77	12.2	12.3	1.83
		MAE	4.38	6.62	5.02	3.93	5.60	2.06	4.63	4.05	6.18	0.77

Day	Weather	Error [%]	Power				
			SDM	SDMnp	DDM	DDMnp	Sandia
22/01/2018	Clear day	RMS	3.65	8.49	3.45	4.34	3.79
		MAE	4.24	6.79	3.74	4.44	4.51
29/01/2018	Semi-cloudy	RMS	2.48	4.30	2.48	4.30	0.82
		MAE	5.14	11.7	4.58	10.3	2.33
28/01/2018	Cloudy day	RMS	5.35	4.82	5.51	5.24	2.79
		MAE	1.37	1.56	1.55	1.57	5.63

Table 10.
Calculated RMS (%) and MAE (%).

where N is the number of samples [41].

The aforementioned results show clearly that the extracted parameters of the PV module using dynamic techniques present more accuracy compared with the static method and the parameters given by the manufacturer. Indeed, the parameters obtained by the static method are clearly improved for variable weather conditions (irradiation and temperature), which is confirmed using different skies. **Table 10** summarizes the calculated RMS and MAE errors values which show obviously that the developed models present advantages comparing with real outdoor data of different weather conditions.

6. Conclusion

In this chapter, both dynamic and static parameters identification methods have been highlighted and compared with real measurement. The SDM and DDM nominal parameters involved in I_0 and I_{ph} equations have been developed by dynamic method. This improved result has been compared with that given by the static technique and Sandia model versus out-door real data for different skies (clear day, semi-cloudy day and cloudy day). It was found that SDM and DDM based on the parameters extracted by dynamic method give satisfactory accuracy, which is confirmed by some calculated indicator such as: the hourly efficiency and both root mean square error (RMS) and the mean absolute error (MAE). This allows to solve modeling problems of PV module that apply for several applications such as fault detection.

Author details

Sid-Ali Blaifi[1,2]* and Bilal Taghezouit[3]

1 Research Laboratory of Electrical Engineering and Automatic LREA, University of Medea, Medea, Algeria

2 Department of Science and Technologies, University of Khemis Miliana, Algeria

3 Centre de Développement des Energies Renouvelables, CDER, Algiers, Algeria

*Address all correspondence to: sidali.blaifi@gmail.com; s.blaifi@univ-dbkm.dz

References

[1] AbdulHadi M, Al-Ibrahim AM, Virk GS. Neuro-fuzzy-based solar cell model. IEEE Transactions on Energy Conversion. September 2004;**19**(3): 619-624

[2] Mekki H, Mellit A, Salhi H, Khaled B. Modeling and simulation of photovoltaic panel based on artificial neural networks and VHDL-language. In: Electronics, Circuits and Systems, 2007. ICECS 2007. 14th IEEE International Conference.

[3] Mellit A, Saglam S, Kalogirou SA. Artificial neural network-based model for estimating the produced power of a photovoltaic module. Renewable Energy. 2013;**60**:71-78

[4] Askarzadeh A. Voltage prediction of a photovoltaic module using artificial neural networks. International Transactions on Electrical Energy Systems. 2014;**24**:1715-1725

[5] Ortiz-Rivera EI, Peng FZ. Analytical Model for a Photovoltaic Module Using the Electrical Characteristics provided by the Manufacturer Data Sheet Power Electronics Specialists Conference. PESC '05. IEEE 36th. 2005

[6] King DL, Boyson WE, Kratochvil JA. Photovoltaic array performance model. Sandia Report No. SAND 2004-3535. 2004. Available from: http://www.sandia.gov/pv/docs/PDF/King%20SAND.pdf

[7] Zhou W, Yang H, Fang Z. A novel model for photovoltaic array performance prediction. Applied Energy. 2007;**84**:1187-1198

[8] Deline C, Dobos A, Janzou S, Meydbray J, Donovan M. A simplified model of uniform shading in large photovoltaic arrays. Solar Energy. 2013; **96**:274-282

[9] Chenni R, Makhlouf M, Kerbache T, Bouzid A. A detailed modeling method for photovoltaic cells. Energy. 2007;**32**: 1724-1730

[10] Rosell JI, Ibanez M. Modelling power output in photovoltaic modules for outdoor operating conditions. Energy Conversion and Management. 2006;**47**:2424-2430

[11] Castaner L, Silvestre S. Front matter. In: Castaner RL, Silvestre S, editors. Modeling Photovoltaic Systems Using PSpice. Chichester, England: Wiley; 2003. p. 376

[12] Ishaque K, Salam Z, Taheri H, Shamsudin A. A critical evaluation of EA computational methods for photovoltaic cell parameter extraction based on two diode model. Solar Energy. 2011;**85**:1768-1779

[13] Easwarakhanthan T, Bottin J, Bouhouch I, Boutrit C. Nonlinear minimization algorithm for determining the solar cell parameters with microcomputers. International Journal of Solar Energy. 1986;**4**:1-12

[14] Dkhichi F, Oukarfi B, Fakkar A, Belbounaguia N. Parameter identification of solar cell model using LevenbergeMarquardt algorithm combined with simulated annealing. Solar Energy. 2014;**110**:781-788

[15] El-Naggar KM, AlRashidi MR, AlHajri MF, Al-Othman AK. Simulated annealing algorithm for photovoltaic parameters identification. Solar Energy. 2012;**86**:266-274

[16] AlHajri MF, El-Naggar KM, AlRashidi MR, Al-Othman AK. Optimal extraction of solar cell parameters using pattern search. Renewable Energy. 2012; **44**:238-245

[17] Chen Z, Wu L, Lin P, Wu Y, Cheng S. Parameters identification of

photovoltaic models using hybrid adaptive Nelder-Mead simplex algorithm based on eagle strategy. Applied Energy. 2016;**182**:47-57

[18] Hamid NFA, Rahim NA, Selvaraj J. Solar cell parameters identification using hybrid Nelder-Mead and modified particle swarm optimization. Journal of Renewable and Sustainable Energy. 2016;**8**. DOI: 10.1063/1.4941791

[19] Jervase JA, Bourdoucen H, Al-Lawati A. Solar cell parameter extraction using genetic algorithms. Measurement Science and Technology. 2001;**12**(11):1922

[20] Huang W, Jiang C, Xue L, Song D. Extracting solar cell model parameters based on chaos particle swarm algorithm. In: 2011 International Conference on Electric Information and Control Engineering. Vol. 2011. pp. 398-402

[21] Meiying Y, Xiaodong W, Yousheng X. Parameter extraction of solar cells using particle swarm optimization. Journal of Applied Physics. 2009;**105**:094502-094508

[22] Hachana O, Hemsas KE, Tina GM, Ventura C. Comparison of different metaheuristic algorithms for parameter identification of photovoltaic cell/module. Journal of Renewable and Sustainable Energy. 2013;**5**:053122

[23] Ma J, Man KL, Guan SU, Ting TO, Wong PWH. Parameter estimation of photovoltaic model via parallel particle swarm optimization algorithm. International Journal of Energy Research. 2016;**40**: 343e352

[24] Jordehi AR. Time varying acceleration coefficients particle swarm optimization (TVACPSO): A new optimization algorithm for estimating parameters of PV cells and modules. Energy Conversion and Management. 2016;**129**:262-274

[25] Muralidharan R. Parameter extraction of solar photovoltaic cells and modules using current-voltage characteristics. International Journal of Ambient Energy (2016). Europe-China cooperation and competition in the Green Industries; 2017;**38**(5):509-513. DOI: 10.1080/01430750. 2016.1144525

[26] Oliva D, Cuevas E, Pajares G. Parameter identification of solar cells using artificial bee colony optimization. Energy. 2014;**72**:93-102

[27] Ma J, Ting TO, Man KL, Zhang N, Guan S-U, Wong PWH. Parameter estimation of photovoltaic models via Cuckoo search, Hindawi Publish. Corporat. Journal of Applied Mathematics. 2013;**362619**. 8 pages

[28] Jamadi M, Merrikh-Bayat F, Bigdeli M. Very accurate parameter estimation of single- and double-diode solar cell models using a modified artificial bee colony algorithm. International Journal of Energy and Environmental Engineering. 2016;**7**: 13-25

[29] Askarzadeh A, Rezazadeh A. Artificial bee swarm optimization algorithm for parameters identification of solar cell models. Applied Energy. 2013;**102**:943-949

[30] Alam DF, Yousri DA, Eteiba MB. Flower pollination algorithm based solar PV parameter estimation. Energy Conversion and Management. 2015;**101**: 410-422

[31] Benkercha R, Moulahoum S, Colak I, Taghezouit B. PV module parameters extraction with maximum power point estimation based on flower pollination algorithm. In: 2016 IEEE International Power Electronics and Motion Control Conference, PEMC. 2016. pp. 442-449

[32] Ram JP, Babu TS, Dragicevic T, Rajasekar N. A new hybrid bee pollinator flower pollination algorithm for solar PV parameter estimation. Energy Conversion and Management. 2017;**135**:463-476

[33] Askarzadeh A, Rezazadeh A. Parameter identification for solar cell models using harmony search-based algorithms. Solar Energy. 2012;**86**: 3241-3249

[34] Han W, Wang H-H, Chen L. Parameters identification for photovoltaic module based on an improved artificial fish swarm algorithm. Scientific World Journal. 2014;**859239**. 12 pages

[35] Kichou S, Silvestre S, Guglielminotti L, et al. Comparison of two PV array models for the simulation of PV systems using five different algorithms for the parameters identification. Renewable Energy. 2016; **99**:270-279

[36] Blaifi S-a, Moulahoum S, Taghezouit B, Saim A. An enhanced dynamic modeling of PV module using Levenberg-Marquardt algorithm. Renewable Energy. 2019;**135**:745-760

[37] Blaifi S, Moulahoum S, Colak I, Merrouche W. An enhanced dynamic modelof battery using genetic algorithm suitable for photovoltaic applications. Applied Energy. 2016;**169**:888-898

[38] Blaifi S, Moulahoum S, Colak I, Merrouche W. Monitoring and enhanced dynamic modeling of battery by genetic algorithm using LabVIEW applied in photovoltaic system. Electrical Engineering. June 2018;**100** (2):1021-1038

[39] Blaifi S, Moulahoum S, Benkercha R, Taghezouit B, Saim A. M5P model tree based fast fuzzy maximum power point tracker. Solar Energy. 2018;**163**:405-442

[40] Benyoucef AS et al. Artificial bee colony based algorithm for maximum power point tracking (MPPT) for PV systems operating under partial shaded conditions. Applied Soft Computing. 2015;**32**:38-48

[41] Blaifi S, Moulahoum S, Colak I, Merrouche W. An enhanced dynamic model of battery using genetic algorithm suitable for photovoltaic applications. Applied Energy. 2016;**169**: 888-898

Chapter 5

Performance Analysis of a 20 MW Grid-Connected Photovoltaic Installation in Adrar, South of Algeria

Nouar Aoun

Abstract

This chapter presents the performance of a 20 MWp grid-connected PV system installed in a harsh environment, Adrar in the South of Algeria. The results were monitored over a period of 1 year, from January 2018 to December 2018. The PV system assessment includes final and reference yields, system efficiency, performance ratio, capacity factor, and total system losses. The total yearly energy delivered to the grid was 35892.22 MWh, and the monthly average reference and final yields of the system are in the range 5.92–8.1 (h/day) and 4.39–5.56 (h/day), respectively. Furthermore, the annual average daily PV system efficiency, performance ratio, capacity factor, and total losses were 10.82%, 71.71%, 20.76%, and 2.04 h/day, respectively.

Keywords: photovoltaic, grid-connected, final yield, PV system efficiency, performance ratio, total losses

1. Introduction

In recent years, Algeria has shown great interest in new and renewable energies. Therefore, on February 19, 2015, the first PV plant was commissioned in Djanet (stand-alone system) in the South-East of Algeria with a total power of 3 MW. In the same year, five PV plants with a total capacity of 48 MW were integrated into the national grid. Most of the photovoltaic plants were installed in the south of Algeria where the insulation time exceeds 3500 h/year. Evaluating the performance of the installed photovoltaic systems is indispensable; it gives researchers the required information regarding the technology (PV, inverter) used and its suitability for the installation site.

In the literature, many studies have investigated, evaluated, and analyzed PV grid-connected systems in different locations, such as India [1], Italy [2], Ireland [3], Ghana [4], Iran [5], Brazil [6], Mauritania [7], Korea [8], Greece [9], and Morocco [10].

In India (South India), the performance of 5 MWp photovoltaic grid-connected systems was presented in [11], by using the RETScreen software for the validation of the performance of the PV plant in real time. It was found that the annual average energy generated by the system is 24116.61 kWh/day, and observed that the annual average daily array yield is of 5.46 h/day, reference yield of 5.128 h/day, final yield of 4.810 h/day, module efficiency of 6.08%, inverter efficiency of 88.20%, and system efficiency of 5.08%. Moreover, a performance study of a 190 kWp grid-interactive photovoltaic plant was conducted by Sharma and Chandel [12], who found that the final yield, reference yield, and performance ratio vary in the range 1.45–2.84 h/day, 2.29–3.53 h/day, and 55–83%, respectively. In [13], the authors presented a grid-connected photovoltaic system installed on the rooftop of the Siksha 'O' Anusandhan University. The results showed the total energy generated during the period test was 14.960 MWh. The array yield, final yield, module efficiency, inverter efficiency, and performance ratio of the system were found to be 4.09 h/day, 3.67 h/day, and 13.42, 89.83, and 78%, respectively. In a different study, Doolla and Banerjee [14] analyzed the variation in the capacity factor of PV grid-connected systems around India. They found that the capacity factor varies between 16 and 20%, and lower values were observed in northeastern states of the country due to poor solar irradiance throughout the year.

In other studies, Ya'acob et al. [15] evaluated three different PV systems in Malaysia, namely concentrating PV system (CPV), PV sun tracking flat (TF), and fixed flat PV system (FF). It was concluded that the most suitable system is the tracking flat PV system with an average daily generation, system efficiency, power efficiency, average daily yield, and capacity factor of 4.7 kWh, 11%, 85%, 2.3 kWh/kWp, and 32%, respectively. Mpholo et al. [16] evaluated the performance of a 281 kWp grid-connected photovoltaic system in Lesotho, Roma (Italy). The results show a satisfactory performance, with a weighted performance ratio of 70%; a monthly yield for the array (YA), reference (YR), and final (YF) in the ranges (71.7–168.0), (161.6–199.3), and (64.4–151.4) kWh/kWp, respectively.

In Algeria, although different photovoltaic systems were installed around the country, they were poorly investigated. Only some efforts were found in the literature to evaluate the performance of mini-systems connected to the power grid [17, 18].

For the first time, this chapter attempts to predict the performance of a grid-connected system installed in a desert area in the Adrar region, South of Algeria, and a performance evaluation of a 20 MW PV system over a 1-year period (January 2018 to December 2018) is presented. The IEA 61724 standard (International Energy agency) guidelines were used to analyze the performance of the system. The PV system assessment includes final and reference yields, system efficiency, performance ratio, capacity factor, and total system losses.

2. PV plant description

The 20 MW grid-connected PV plant of Adrar comprises 20 subfields of 1 MWc each and divided into two subsystems. In total, the numbers of modules per subfield are 4092. The modules are mounted at an angle of 28° and oriented southward. The type of module is polycrystalline with a maximum power of 245 Wp and total module area of 132,538,775 m^2. The manufacture characteristic of the PV module is shown in **Table 1**. The station was put into service on October 28, 2015, and was expected to produce annually 36,414 MWh, with an economy of about 10,080 m^3/year of natural gas. However, the results presented in this study consist of the measured data over a period of 1 year (January 2018 to December 2018). A block scheme of the considered photovoltaic system is illustrated in **Figure 1**.

The in-plane global solar radiation, wind speed, ambient temperature, and the output AC power were measured and recorded after every 15-min interval.

Parameters	Specification
Manufacturer	Yingli Green Energy Holding Co. Ltd.
Cell type	Polycristallin
PV Model	YL245P-29P
Maximum power, P_p (W)	245
Maximum power voltage, V_p (V)	29.6
Maximum power current, I_p (A)	8.28
Open circuit voltage, V_{oc} (V)	37.5
Short circuit current, I_{sc} (A)	8.83
Number of cells	60
Module dimensions (mm)	1640*990*35
Module efficiency (%)	15.1
Maximum system voltage (V)	1000
Temperature coefficient of P_p, α_{Pp} (%/°C)	−0.42
Temperature coefficient of I_{sc}, α_{Isc} (%/°C)	0.05
Temperature coefficient of V_{oc}, α_{Voc} (%/°C)	−0.32

Table 1.
Photovoltaic module specifications.

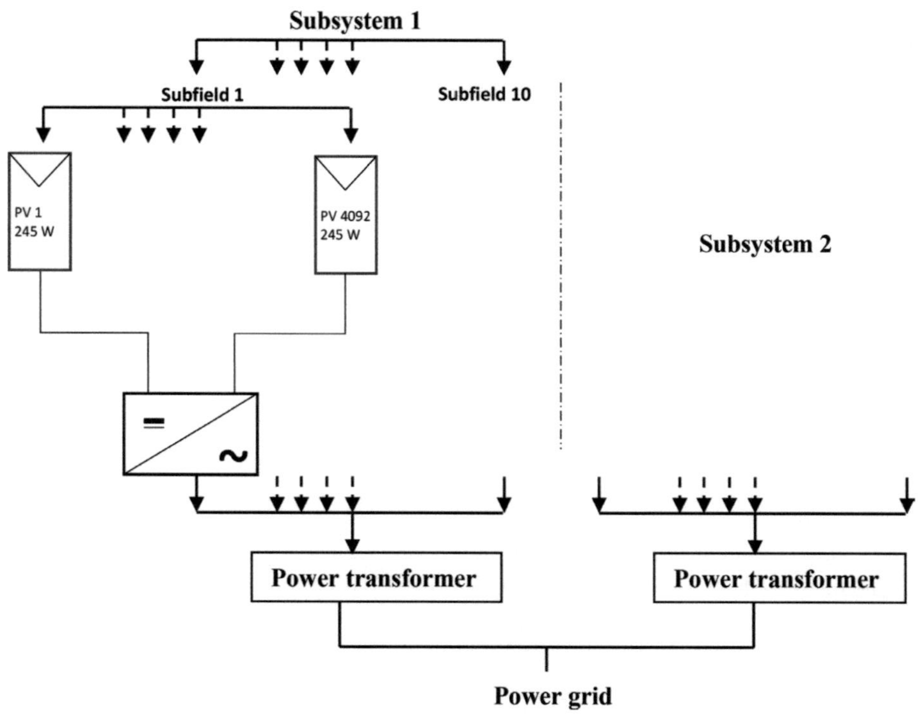

Figure 1.
Schematic representation of the considered PV system.

We observe a few days missing in the months of: January (2-day data is missing for both subfields), July (5-day data is missed for the second subsystem), and October (3 days are missed concerning the two subsystem).

3. Performance analysis

Equations (1)–(9) are applied in accordance with the IEC 61724 standard to investigate the performance of the grid-connected PV system.

3.1 Final yield (Y_F)

Final yield (Y_F) is the total AC energy generated by the PV system over the monitored period divided by the rated output power of the PV system:

$$Y_{F,d} = \frac{E_{AC,d}}{P_{PV,rated}} \tag{1}$$

$$Y_{F,m} = \frac{1}{N}\sum_{1}^{N} Y_{F,d} \tag{2}$$

The total daily and monthly AC energy generated by the system was obtained as:

$$E_{AC,d} = \tau \cdot \sum_{1}^{24} P_p \tag{3}$$

$$E_{AC,m} = \sum_{1}^{N} E_{AC,d} \tag{4}$$

where N is the number of days per month, τ is the recording interval (τ = 15mn), P_p is the AC power in kW, and E_{AC} is expressed in kWh. The terms *d* and *m* indicate the daily and monthly period.

3.2 Reference yield (Y_R)

Reference yield is the ratio of total in-plane solar irradiation H_t (kW/m^2) and the reference irradiance H_{STC} (=1 kW/m^2). It can be calculated as follows:

$$Y_{R,d} = \frac{\tau \cdot \sum G_t}{G_{STC}} \tag{5}$$

The monthly average daily reference yield is obtained as:

$$Y_{R,m} = \frac{1}{N}\sum_{1}^{N} Y_{R,d} \tag{6}$$

3.3 Performance ratio (PR)

The performance ratio is a dimensionless quantity which provides important information concerning the system losses (modules, inverters, cables, weather conditions, loss due to non-STC temperature) in the PV systems. This quantity defined as the ratio of the final yield (Y_F) and the reference yield (Y_R). It indicates the percentage of the real energy supplied by the system [19, 20] and can be expressed as follows:

$$PR = \frac{Y_F}{Y_R} \tag{7}$$

3.4 Total energy losses (L_T)

The total energy loss (T_L) of the system is obtained from the difference between the reference yield Y_R and the final yield Y_F. The total losses represent the PV losses due to irradiance level and array temperature and quality module, ohmic wiring, mismatch, and total inverter losses [21]:

$$L_T = Y_R - Y_F \tag{8}$$

3.5 Capacity factor (CF)

The capacity factor (CF) is the ratio of the actual energy output of the PV system and the PV system energy generates operating at full rated power. The yearly and monthly capacity factors were calculated by Eqs. (9) and (10), respectively:

$$CF_y = \frac{E_{AC,y}}{P_{PV,rated} \times 8760} \tag{9}$$

$$CF_m = \frac{E_{AC,m}}{P_{PV,rated} \times 24 * N} \tag{10}$$

where the term (y) indicate the yearly period.

3.6 System efficiency

The monthly system efficiency is given as:

$$\eta_{sys,m} = \frac{E_{AC,m}}{H_{t,m} \times A_a} \tag{11}$$

where

$$H_{t,m} = \sum_1^N \left(\tau \sum_1^{24} G_t \right) \tag{12}$$

where $H_{t,m}$ monthly total in-plane insolation (kWh/m^2).

4. Results and discussion

The monthly average daily ambient temperature and wind speed over the monitored period is shown in **Figure 2**. The monthly average daily ambient temperature varies from a minimum of 14.05°C in December to a maximum of 40.8°C in July. The months of May, June, July, August, and September (summer months) show higher ambient temperature compared to the other months. The monthly variation of daily average wind speed is from a minimum of 3.59 m/s in October to a maximum of 5.14 m/s in April. **Figure 2** shows that the wind speed over the monitored period (during all the months) is higher.

The monthly total AC energy generated to the grid and in-plane irradiation over the monitored period is shown in **Figure 3**. The total in-plane irradiation varies from a minimum of 171.7 kWh/m^2 in January to a maximum of 251.2 kWh/m^2 in May. The highest values were recorded in the months of March, April, May, June, July, and August (Spring and Summer). The monthly total variation of the AC energy generated to the grid ranged from 2725.15 MWh in July to 3447.5 MWh in March. As shown in **Figure 3**, there is a correlation between the total AC energy and the total in-plane irradiation. As the total irradiation increases, the total energy likewise increases. In July, a 5-day loss of energy value data for one of the subsystem was observed. This explains the low energy value in the month of July despite high irradiation. From **Figures 2** and **3**, it is evidently clear that increases in ambient

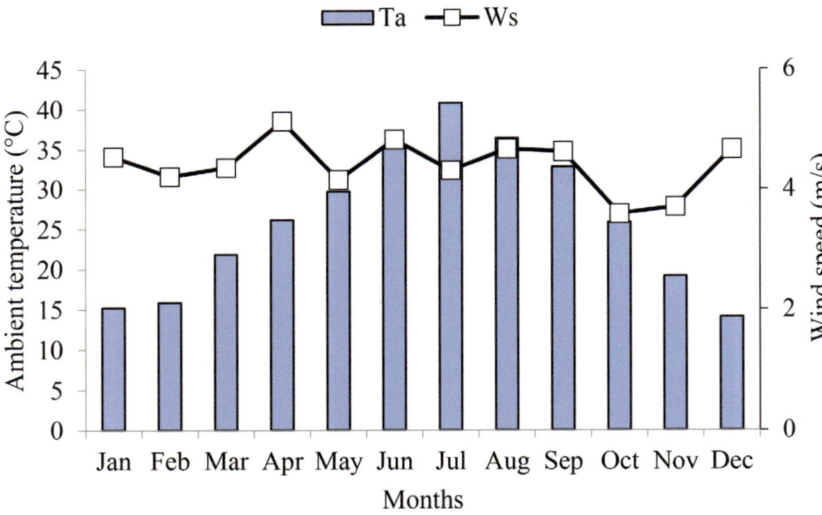

Figure 2.
Monthly average daily ambient temperature and wind speed over the monitored period.

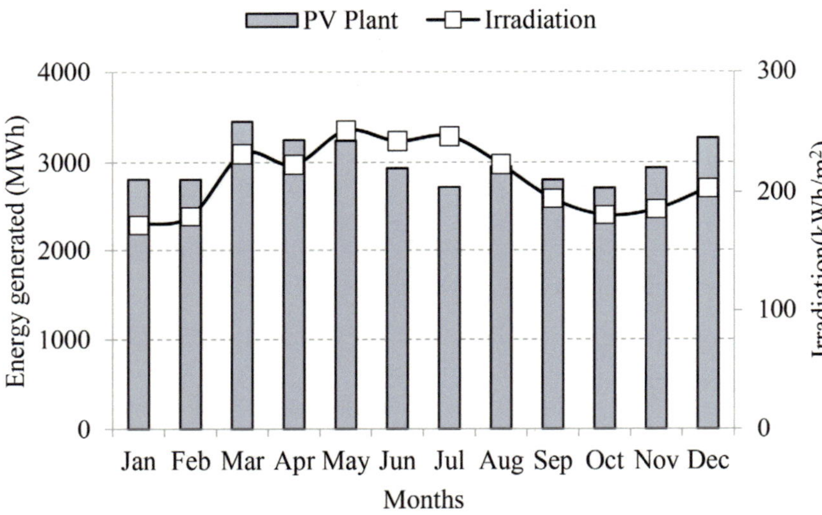

Figure 3.
Monthly total AC energy generated to the grid and in-plane irradiation over the monitored period.

temperature have a negative effect on the energy produced by the system due to the increased PV modules temperature, although the wind flow on the photovoltaic modules contributes to reducing the PV cell temperature [7, 19]. Generally, the irradiation had a greater impact on energy generation than ambient temperature [22].

Figure 4 shows the average monthly performance ratio and efficiency of the PV system over the monitored period. The results show that the performance ratio varies from a min. of 55.29% in July to a maximum of 81.96% in January, with an annual average of 71.71%. Similarly, the efficiency varies from 8.34% in July to 12.37% in January. The performance ratio and efficiency refer to the level of losses in the photovoltaic solar system. There are three main reasons for losses of performance are dust, ambient temperature, and solar radiation. In [22], the authors indicated that ambient temperature had a greater impact on the performance ratio and efficiency than irradiation, and high values for both parameters were observed during days with low ambient temperature.

Figure 5 shows the monthly and annual capacity factor of the PV system. The monthly capacity factor can be seen to vary between 18.31% in July and 23.17% in March. The annual capacity factor is 20.77%, with an overall monthly average of 20.76%. The capacity factor of other PV plants based on existing literature: in India [1], it ranged from 15.4 to 20%, Oman from 13 to 20% [23] and 21% [24], and Morocco from 6.55 to 21.42% in [10] and for a polycrystalline PV technologies the CF ranged from 20.014 to 23.799% [25].

Figure 6 shows the monthly daily average variation of the final yield, reference yield, and total energy losses. As shown in **Figures 3** and **6**, the reference yield is directly proportional to the irradiation, respectively; they vary from a minimum of 5.92 (h/day) and 171.69 kWh/m^2 in January to a maximum of 8.1 (h/day) and 251.19 kWh/m^2 in May.

In the present study, the annual monthly average daily final yield of the PV plant is 4.98 (h/day). The reason for this high value is the high irradiation and the extended

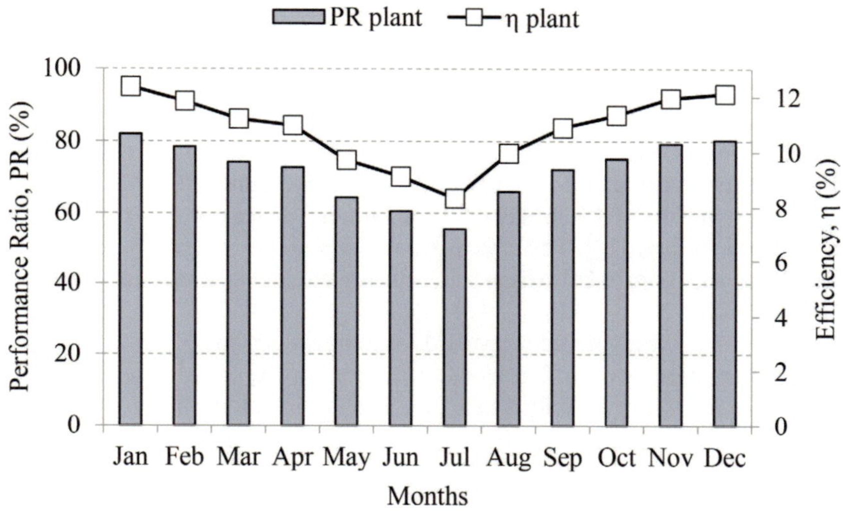

Figure 4.
Performance ratio and efficiency of the PV system over the monitored period.

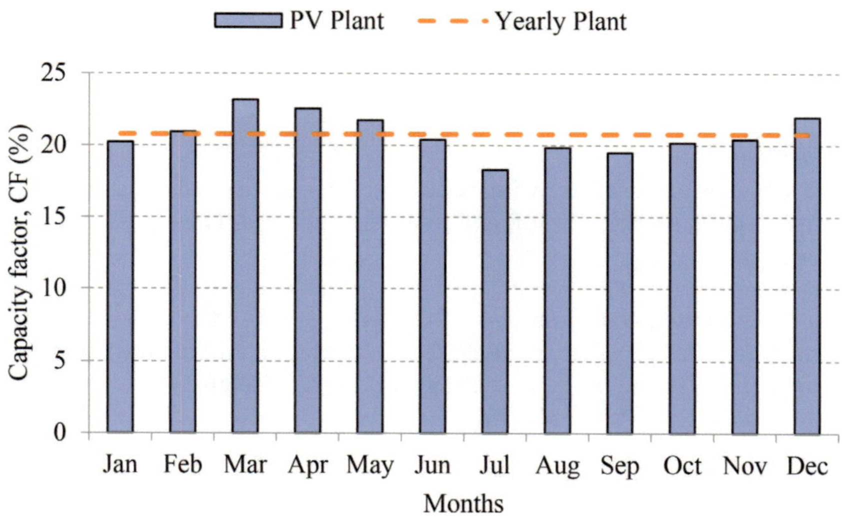

Figure 5.
Monthly and annual capacity factor over the monitored period.

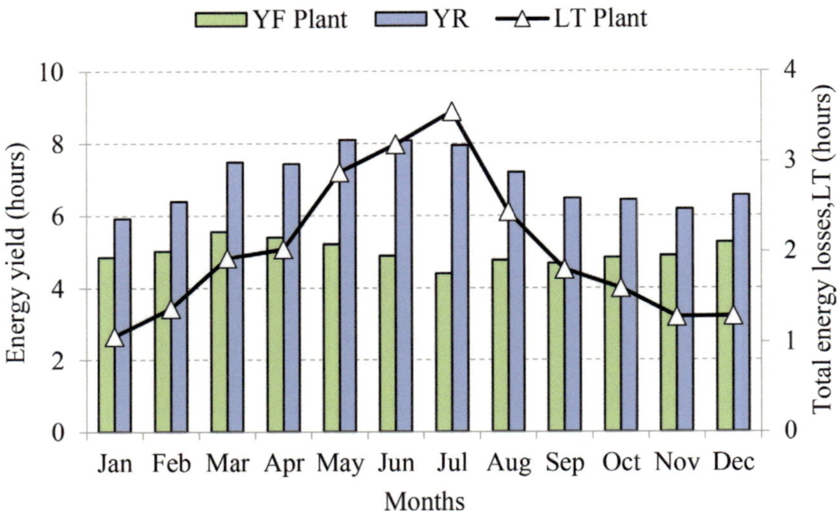

Figure 6.
Monthly average energy yield and energy losses over the monitored period.

sunshine duration in Adrar [10]. This value is close to other final yields of different PV systems found in the literature such as that of Oman (Sohar) 5.1 (h/day) [24], and higher than in Kuwait 4.5 (h/day) [26], in Oman (Muscat) 4.1 (h/day) [23] and in Morocco 4.45 (h/day) [10]. The monthly daily average yields (final and reference) increase during the months from May to August due to higher irradiation and more sun hours.

The total energy losses varied between 1.06 h in January and 3.55 h in July. The higher total energy losses were observed in the hot months, i.e., May to August. The reasons for the losses are generally due to dust and ambient temperature (heating of the photovoltaic cells which reduce the power provided by the photovoltaic modules). And in these conditions, periods of downtime of the system were observed (2 days in January, 5 days in July and in October stopped for 3 days with 1-day data incomplete).

5. Conclusion

The present chapter investigates the performance analysis assessment of ground-mounted large-scale grid-connected PV plant in Adrar, South of Algeria. The PV system is made up of 20 MW of polycrystalline silicon modules produced by Yingli Green Energy Holding Co. Ltd. and monitored between January 2018 and December 2018.

The PV system assessment includes final and reference yields, system efficiency, performance ratio, capacity factor, and total system losses. The main findings are the followings:

- The monthly total variation of the AC energy generated to the grid ranged from 2725.15 MWh in July to 3447.5 MWh in March, and the total yearly energy delivered to the grid in the year 2018 was 35892.22 MWh, which is close to the expected energy (36,414 MWh).

- The monthly average reference and final yields of the system are in the range 5.92–8.1 (h/day) and 4.39–5.56 (h/day), respectively, and the annual monthly average daily final yield is 4.98 (h/day). The monthly daily average yields (final and reference) increase from May to August due to higher irradiation and extended sunshine duration.

- The minimum and maximum efficiency and performance ratio were 8.34 and 55.29% in July, and 12.37 and 81.96% in January, respectively. The minimum capacity factor was 18.31 in July and the maximum was 23.17% in March. The minimum total energy losses was 1.07 h in January and the maximum was 3.55 h in July. Furthermore, the annual average daily PV system efficiency, performance ratio, capacity factor, and total losses were 10.82%, 71.71%, 20.76%, and 2.04 h/day, respectively.

- The reasons for the losses are generally due to dust, ambient temperature and solar radiation.

The results obtained show that the installation of photovoltaic stations in Adrar in Southern Algeria gives good results encouraging investments in this field and region.

Conflict of interest

The author has no conflict of interest statement.

Appendices and nomenclature

P_{mp}	maximum power (W)
V_{mp}	maximum power voltage (V)
I_{mp}	maximum power current (A)
V_{oc}	open circuit voltage (V)
I_{sc}	short circuit current (A)
α_{Pmp}	temperature coefficient of Pmp (%/°C)
α_{Isc}	temperature coefficient of Isc (%/°C)
α_{Voc}	temperature coefficient of Voc (%/°C)
STC	standard test condition (25°C and 1 kW/m^2)
T_a	ambient temperature (°C)
W_s	wind speed (m/s)
P_p	AC power (kW)
E_{AC}	AC energy generated (kWh)
Y_F	final yield (h)
Y_R	reference yield (h)
T_L	total energy losses (h)
CF	capacity factor (%)
PR	performance ratio (%)
H_t	total in-plane solar irradiation (kW/m^2)
H_{STC}	reference irradiation (=1 kW/m^2)
η_{sys}	system efficiency (%)
N	number of days per month
d, m, and y	daily, monthly, and yearly period

Author details

Nouar Aoun
Unité de Recherche en Energies Renouvelables en Milieu Saharien, URERMS, Centre de Développement des Energies Renouvelables, Adrar, Algeria

*Address all correspondence to: nouar.aoun@gmail.com

References

[1] Kumar NM, Gupta RP, Mathew M, Jayakumar A, Singh NK. Performance, energy loss, and degradation prediction of roof-integrated crystalline solar PV system installed in northern India. Case Studies in Thermal Engineering. 2019;**13**:100409. DOI: 10.1016/j.csite.2019.100409

[2] Malvoni M, Leggieri A, Maggiotto G, Congedo PM, De Giorgi MG. Long term performance, losses and efficiency analysis of a 960 kWP photovoltaic system in the Mediterranean climate. Energy Conversion and Management. 2017;**145**:169-181. DOI: 10.1016/j.enconman.2017.04.075

[3] Ayompe LM, Duffy A, McCormack SJ, Conlon M. Measured performance of a 1.72 kW rooftop grid connected photovoltaic system in Ireland. Energy Conversion and Management. 2011;**52**:816-825. DOI: 10.1016/j.enconman.2010.08.007

[4] Quansah DA, Adaramola MS, Appiah GK, Edwin IA. Performance analysis of different grid-connected solar photovoltaic (PV) system technologies with combined capacity of 20 kW located in humid tropical climate. International Journal of Hydrogen Energy. 2017;**42**:4626-4635. DOI: 10.1016/j.ijhydene.2016.10.119

[5] Edalati S, Ameri M, Iranmanesh M, Tarmahi H, Gholampour M. Technical and economic assessments of grid-connected photovoltaic power plants: Iran case study. Energy. 2016;**114**:923-934. DOI: 10.1016/j.energy.2016.08.041

[6] de Lima LC, de Araújo Ferreira L, de Lima Morais FHB. Performance analysis of a grid connected photovoltaic system in northeastern Brazil. Energy for Sustainable Development. 2017;**37**:79-85. DOI: 10.1016/j.esd.2017.01.004

[7] Elhadj Sidi CEB, Ndiaye ML, El Bah M, Mbodji A, Ndiaye A, Ndiaye PA. Performance analysis of the first large-scale (15 MWp) grid-connected photovoltaic plant in Mauritania. Energy Conversion and Management. 2016;**119**:411-421. DOI: 10.1016/j.enconman.2016.04.070

[8] Kim JY, Jeon GY, Hong WH. The performance and economical analysis of grid-connected photovoltaic systems in Daegu, Korea. Applied Energy. 2009;**86**:265-272. DOI: 10.1016/j.apenergy.2008.04.006

[9] Kymakis E, Kalykakis S, Papazoglou TM. Performance analysis of a grid connected photovoltaic park on the island of Crete. Energy Conversion and Management. 2009;**50**:433-438. DOI: 10.1016/j.enconman.2008.12.009

[10] Attari K, Elyaakoubi A, Asselman A. Performance analysis and investigation of a grid-connected photovoltaic installation in Morocco. Energy Reports. 2016;**2**:261-266. DOI: 10.1016/j.egyr.2016.10.004

[11] Sundaram S, Babu JSC. Performance evaluation and validation of 5 MWp grid connected solar photovoltaic plant in South India. Energy Conversion and Management. 2015;**100**:429-439. DOI: 10.1016/j.enconman.2015.04.069

[12] Sharma V, Chandel SS. Performance analysis of a 190 kWp grid interactive solar photovoltaic power plant in India. Energy. 2013;**55**:476-485. DOI: 10.1016/j.energy.2013.03.075

[13] Sharma R, Goel S. Performance analysis of a 11.2 kWp roof top grid-connected PV system in eastern India. Energy Reports. 2017;**3**:76-84. DOI: 10.1016/j.egyr.2017.05.001

[14] Doolla S, Banerjee R. Diffusion of grid-connected PV in India: An analysis of variations in capacity factor. In: Proceedings of the 35th IEEE

Photovoltaic Specialists Conference; 20-25 June 2010; Honolulu: USA: IEEE; 2010. pp. 002349-002352

[15] Ya'acob ME, Hizam H, Khatib T, Radzi MAM. A comparative study of three types of grid connected photovoltaic systems based on actual performance. Energy Conversion and Management. 2014;**78**:8-13. DOI: 10.1016/j.enconman.2013.10.064

[16] Mpholo M, Nchaba T, Monese M. Yield and performance analysis of the first grid-connected solar farm at Moshoeshoe I international airport. Lesotho - Renewable Energy. 2015;**81**:845-852. DOI: 10.1016/j.renene.2015.04.001

[17] Benatiallah A, Mostefaou R, Bradja K. Performance of photovoltaic solar system in Algeria. Desalination. 2007;**209**:39-42. DOI: 10.1016/j.desal.2007.04.006

[18] Cherfa F, Arab AH, Oussaid R, Abdeladim K, Bouchakour S. Performance analysis of the mini-grid connected photovoltaic system at Algiers. Energy Procedia. 2015;**83**:226-236. DOI: 10.1016/j.egypro.2015.12.177

[19] Dobaria B, Pandya M, Aware M. Analytical assessment of 5.05 kWp grid tied photovoltaic plant performance on the system level in a composite climate of western India. Energy. 2016;**111**:47-51. DOI: 10.1016/j.energy.2016.05.082

[20] Congedo PM, Malvoni M, Mele M, De Giorgi MG. Performance measurements of monocrystalline silicon PV modules in South-Eastern Italy. Energy Conversion and Management. 2013;**68**:1-10. DOI: 10.1016/j.enconman.2012.12.017

[21] Allouhi A, Saadani R, Kousksou T, Saidur R, Jamil A, Rahmoune M. Grid-connected PV systems installed on institutional buildings: Technology comparison, energy analysis and economic performance. Energy and Buildings. 2016;**130**:188-201. DOI: 10.1016/j.enbuild.2016.08.054

[22] Aoun N, Bouchouicha K, Chenni R. Performance evaluation of a mono-crystalline photovoltaic module under different weather and sky conditions. International Journal of Renewable Energy Research. 2017;7:292-297

[23] Al-Badi AH. Measured performance evaluation of a 1.4 kW grid connected desert type PV in Oman. Energy for Sustainable Development. 2018;**47**:107-113. DOI: 10.1016/j.esd.2018.09.007

[24] Kazem HA, Khatib T, Sopian K, Elmenreich W. Performance and feasibility assessment of a 1.4 kW roof top grid-connected photovoltaic power system under desertic weather conditions. Energy and Buildings. 2014;**82**:123-129. DOI: 10.1016/j.enbuild.2014.06.048

[25] Allouhi A, Saadani R, Buker MS, Kousksou T, Jamil A, Rahmoune M. Energetic, economic and environmental (3E) analyses and LCOE estimation of three technologies of PV grid-connected systems under different climates. Solar Energy. 2019;**178**:25-36. DOI: 10.1016/j.solener.2018.11.060

[26] Al-Otaibi A, Al-Qattan A, Fairouz F, Al-Mulla A. Performance evaluation of photovoltaic systems on Kuwaiti schools' rooftop. Energy Conversion and Management. 2015;**95**:110-119. DOI: 10.1016/j.enconman.2015.02.039

Chapter 6

Modeling of the Small Wind Energy in Saharan Region of Algeria

Harrouz Abdelkader, Fadila Tahiri, Boussaid Brahim and Fatiha Bekraoui

Abstract

In the last century, research for the wind power industry started to gain importance in the field of renewable energies. This research is mainly on the side of big wind power. This wind form, are grouped and connected to the electricity grid. In the other form, the small wind power for production in isolated areas. This wind are applicated for example: telecommunication stations, sailboats and isolated areas. This power is from 100 W to a few tens of kilowatts. They are applied to power installations, for example, telecommunication stations, sailboats, and isolated areas. The use of small wind power in the region of Adrar—the southwest of Algeria—is an economical and durable solution. This chapter will focus on the modeling of the main components of a small wind turbine adapted to Saharan regions. The wind chain consists of a wing coupled directly to a synchronous generator that delivers on a continuous bus via a rectifier; it is the overall structure of the chain that we retain for modeling in this chapter. In order to control the system, the modeling of this study touches all the parts of the system: the turbine, the generator PMSG, and the converter with the load. At the end, simulation results are presented to show the good performance of the choice of control type that is applied to the wind system.

Keywords: wind, modeling, electricity, MPPT

1. Introduction

Wind energy is one of the fastest-growing renewable energy in the world. The wind resource comes from the movement of air masses that is directly related to the sunshine of the earth. By warming certain areas of the planet and cooling others a pressure difference is created and the air masses are in perpetual motion.

Since the use of the windmill, wind sensor technology has been evolving. It was in the early 1940s that real prototype wind turbine blades were successfully used to generate electricity. Several technologies are used to capture wind energy (vertical axis or horizontal axis sensor) and the sensor components, the mechanical characteristics of the wind turbine, and the synchronous and asynchronous machines are efficient [1].

In the region of Adrar (located in the southwest of Algeria), there are isolated sites, where the use of this abundant source of energy, is an economical and durable solution. The permanent magnet synchronous machine is characterized by high volumetric torque, very low inertia and low inductances. All of these features

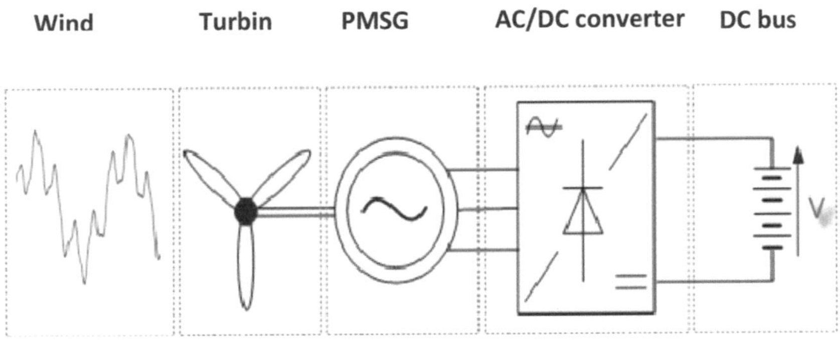

Figure 1.
Schematic diagram of a wind chain.

provide the generator with high performance, high efficiency and better control. This puts this machine in competition with the asynchronous machine [2].

To meet the demand for autonomous electric charge at remote sites, small wind turbines smaller than 5 kW have become a very important system for the production of electricity. Current research will study the functional analysis, control and modeling techniques appropriate for this small turbine system [3].

Modeling of the wind turbine recently published by many authors, Harrouz et al. [4] have modeling of small wind energy based on PMSG in the south of Algeria. Langlois [5] have modeling and study the performances of wind system out the link. Tran [6] has present work of optimal design integrate of wind energy "passive" robustness analysis, experimental validation. Jedli and Hidouri [7] present a power drive scheme for an Isolated pitched wind turbine water pumping system based on DC machine, this work can be a good application of win system in Sahara region. Modeling, simulation, and control of a variable speed wind turbine based on the asynchronous dual-feed generator [8].

The wind energy conversion chain is schematized as shown in **Figure 1**.

This consists of a wing coupled directly to a synchronous generator that delivers on a DC bus via a diode rectifier; it is the structure that we retain for this work.

The control strategies of this system and their possible load connection interfaces must make it possible to capture as much energy as possible over the widest possible range of wind speeds, with the aim of improving the profitability of wind turbine installations.

The section after will focus on the modeling of the main components of a small wind turbine adapted to saharan regions such as Adrar.

2. Modeling of the small wind

2.1 Modeling the wind

The wind is highly variable, both geographically and over time. It varies from one place to another, from 1 day to another, and from 1 s to the next. It is the energy input vector of a wind chain, determining for the calculation of electricity production, therefore for profitability [2].

The dynamic properties of the wind are crucial for the study of the whole system of energy conversion because the wind power, in the optimal conditions, is at the cube of the wind speed (see formula (2)) These fluctuations are influenced by the displacement of air masses at altitude, but also by the relief, the type of soil cover and the thermal stability of the atmosphere. The wind varies in direction and intensity, but for the purposes of the model, it is limited to the variation of the wind in intensity, in a single direction in order to determine a valid and representative sequence of real winds [2–8].

Different approaches used in the literature for the generation of a synthetic series of wind, in our case, the wind speed will be modeled by a sum of several harmonics [2, 4, 9]

$$V_{vent}(t) = A + \sum_{k=1}^{i} a_k \sin(\omega_k . t) \quad (1)$$

where a_k: amplitude of the harmonics; ω_k: frequency of harmonics.

The wind speed is represented by the function:

$$V(t) = 10 + 0.2\sin(0.1047t) + 2\sin(0.2665t) + \sin(1.2930t) + 0.2\sin(3.6645t) \quad (2)$$

It should be noted that this particular wind profile corresponds to [9–12].

2.2 Modeling of turbine

A wind turbine is a machine that, by definition, transforms wind energy into mechanical energy. To begin, it is necessary to quantify the energy source available, i.e., the energy associated with the wind. If the wind has a certain speed "V" at a given moment and crosses a certain area "A," the instantaneous wind power is given by the following relation

$$P_m = \frac{1}{2} C_p \rho A V^3 \quad (3)$$

where ρ is the density of the air, which is approximately 1.2 kg/m^3; A is the area swept by the turbine in m^2; and V is the wind speed in m/s. C_p (power coefficient).

The turbine used in our work is a "Savonius" wind turbine with a vertical axis (see **Figure 2**).

Surface "A" given by the following formula:

$$A = 2R.H \quad (4)$$

where R represents the radius of the wing (m) and H its height (m).

The output power is given by the following equation that we will normalize in pu:

$$P_m = C_p(\lambda).\rho.H.R.V^3 \quad (5)$$

where C_p is a coefficient that expresses the efficiency of the wing in the transformation of the kinetic energy of the wind into mechanical energy, which is in fact often

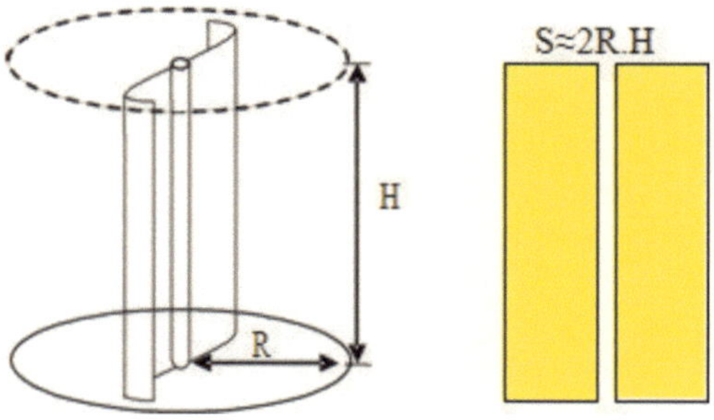

Figure 2.
Wind vertical axis of Savonius type [7].

given as a function of the reduced speed (the specific speed λ), which is the ratio of the linear speed at the end of turbine blades reduced to wind speed, defined by:

$$\lambda = (R.\Omega)/V \quad (6)$$

where Ω the angular rotation speed of the blades in rad/s.

The evaluation of the power coefficient is a data specific to each wind turbine. From the readings taken on a wind turbine, the expression of the power coefficient has been approximated, for this turbine [13, 14], by the analytical equation as a function of λ (resulting from the interpolation) according to:

$$C_p(\lambda) = -0.12992\ \lambda^3 - 0.11681\lambda^2 + 0.45406\ \lambda \quad (7)$$

Figure 3 represents the characteristics of power coefficient as a function of λ, it presents a maximum power factor C_{pmax} = 0.19 for λ_{max} = 0.55.

The wind torque noted C_e is obtained from Eq. (5) and using the expression (6) of the reduced speed (**Figure 4**):

$$C_e = \frac{P_m}{\Omega} = \frac{Cp(\lambda).\rho.R^2.H.V^2}{\lambda} \quad (8)$$

The Model of the wind torque under Simulink presented in **Figure 4**. The **Figure 5** shows the characteristic of the turbine through the wind power extraction of the turbine as a function of the wind speed.

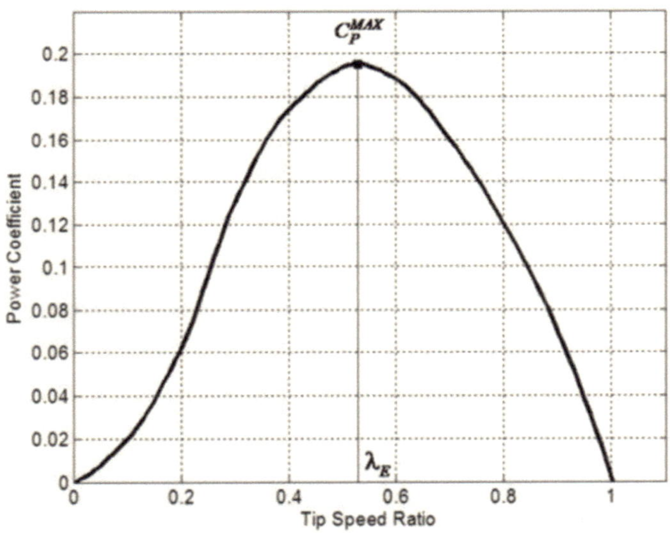

Figure 3.
Power coefficient C_p as a function of λ for Savonius wind turbine.

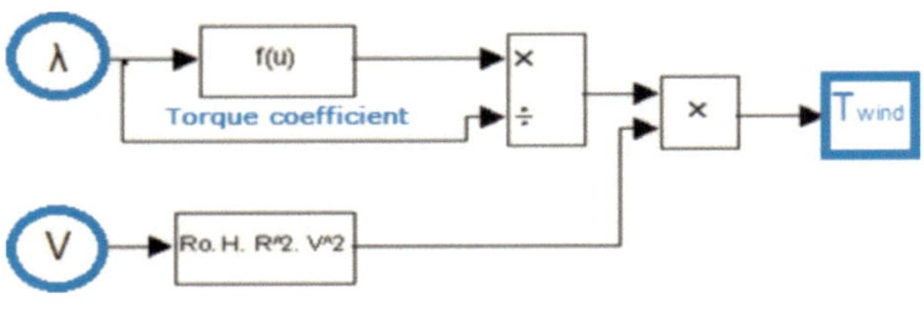

Figure 4.
Model of the wind under Simulink.

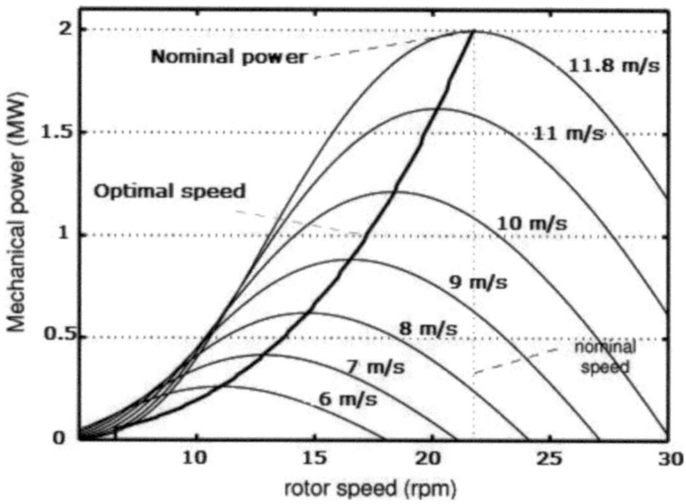

Figure 5.
The output power for different values of wind speed (m/s).

It can be seen from **Figure 5** that the electric power varies with the variation of the mechanical angle of the turbine. The shape of the electric power as a function of the mechanical angle is seems to be that of the Cp as a function of λ. We also note that the higher the wind speed, the greater the power delivered because the power is expressed by the cubic wind speed.

2.3 Mechanical modeling of the turbine-generator coupling

The dynamic (mechanical) behavior of the turbine and generator assembly can be represented by the following relation:

$$J\frac{d\Omega}{dt} = C_e - C_{em} - C_f \tag{9}$$

where Ω the speed on the generator shaft, C_f the friction torque, C_{em} the electromagnetic torque developed by the generator, C_e the mechanical torque applied to the alternator shaft and J is the total moment of inertia, calculated with:

$$J = J_{gen} + \frac{J_t}{i^2} \tag{10}$$

We take note that:

$$C_f = f_m.\Omega \tag{11}$$

where f_m is the coefficient of viscous friction in N.m.

The wind torque provided by the turbine, drives the generator. Taking into account the torque drops caused by the friction and the inertia of the turbine, we can establish the electromechanical model of **Figure 6**.

2.4 Model of the GSAP

According to Chapter, the Permanent Magnet Synchronous Generator (PMSG) is classically modeled in the Park coordinate system, giving rise to the following equation:

$$\begin{cases} V_d = -R_s.I_d - L_d\dfrac{dI_d}{dt} + L_q\omega.I_q \\ V_q = -R_s.I_q - L_q\dfrac{dI_q}{dt} - L_d\omega.I_d + \varphi_f\omega \\ J\dfrac{d\Omega}{dt} = C_m - C_{em} - F.\Omega \\ C_{em} = \dfrac{3}{2}P\left[(Lq - Ld)I_d.I_q + \varphi_f.I_q\right] \end{cases} \tag{12}$$

where: θ is the angle between a reference axis of the stator and an axis of the north pole of the rotor, p-the number of pairs of poles, R_s the resistance of a stator phase.

Also, φ_f is the excitation flux produced by the permanent magnets, L_d and L_q are the equivalent inductances on the d axes, respectively q.

C_{em} is the electromagnetic torque, C_m is the motor torque applied to the generator with F the coefficient of friction, J: moment of inertia and p the number of pairs of poles.

In order to know the dynamic characteristics of the various state variables of the machine and to validate the mathematical model, we proceed to a simulation of the machine using the Runge–Kutta algorithm of order 4.

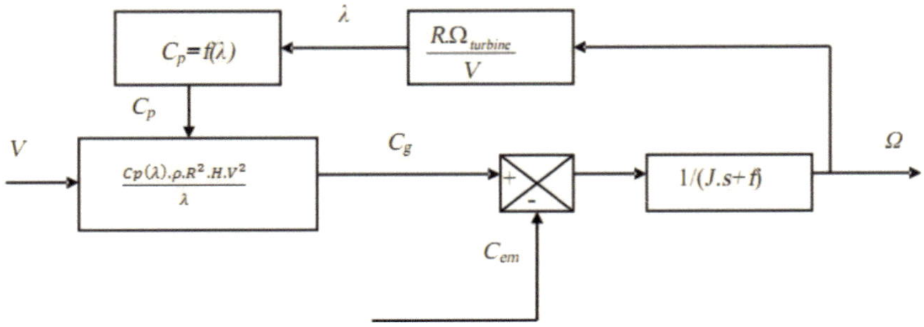

Figure 6.
Model of wind turbine torque.

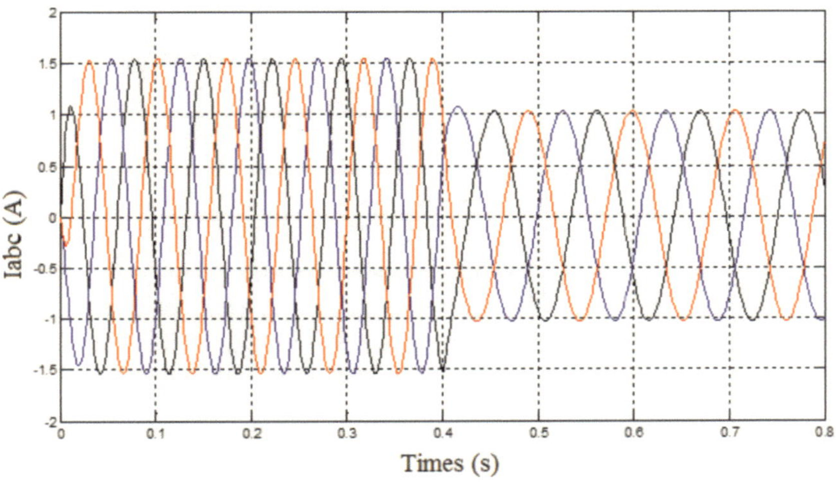

Figure 7.
Stator current I_{abc} of PMSG after simulation.

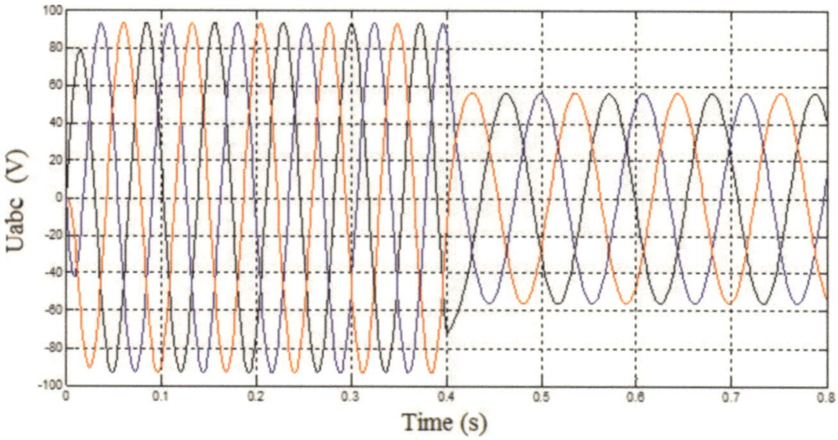

Figure 8.
Stator voltage of PMSG after simulation.

The response to an empty speed step is very fast. The torque peaks at startup and is proportional to the stator current. At full speed, the speed decreases while the torque increases to drive the load (**Figures** 7 and **8**).

To see the efficiency of our model, the next part will be the subject of test on matlab/simulink with real parameters of the wind turbine and the wind speed.

3. Simulation of model PMSG with wind system

The wind will be modeled in deterministic form by a sum of several harmonics:

$$V(t) = 10 + 0.2\sin(0.1047t) + 2\sin(0.2665t) + \sin(1.2930t) + 0.2\sin(3.6645t) \tag{13}$$

The simulation results of the permanent magnet synchronous generator associated with the wind turbine and with the wind speed simulated by Eq. (13) are given by **Figure 9**:

We notice that the results take the form of the wind, the coefficient of power is 0.13 (it is the maximum value to have a maximum of power extracted from the wind). The active power is of the order of 400 W.

But this system presents fluctuations due to the variation of the speed of the wind for that the system needs a command to maintain this power at its nominal value. The next partie present the proposed control of system.

4. Command of wind system

The wind system studied is a complex system composed of several subsystems belonging to several different physical domains, namely, traditional mechanics, electrical engineering and electrochemistry. Energy from the wind passes through the wind turbine that is coupled to the GSAP generator and allows the transformation of mechanical energy into electrical energy. The electrical components such as the static converter disposed downstream of the generator, have a role of active adaptation of the characteristics of the electrical energy between the generator and the final load.

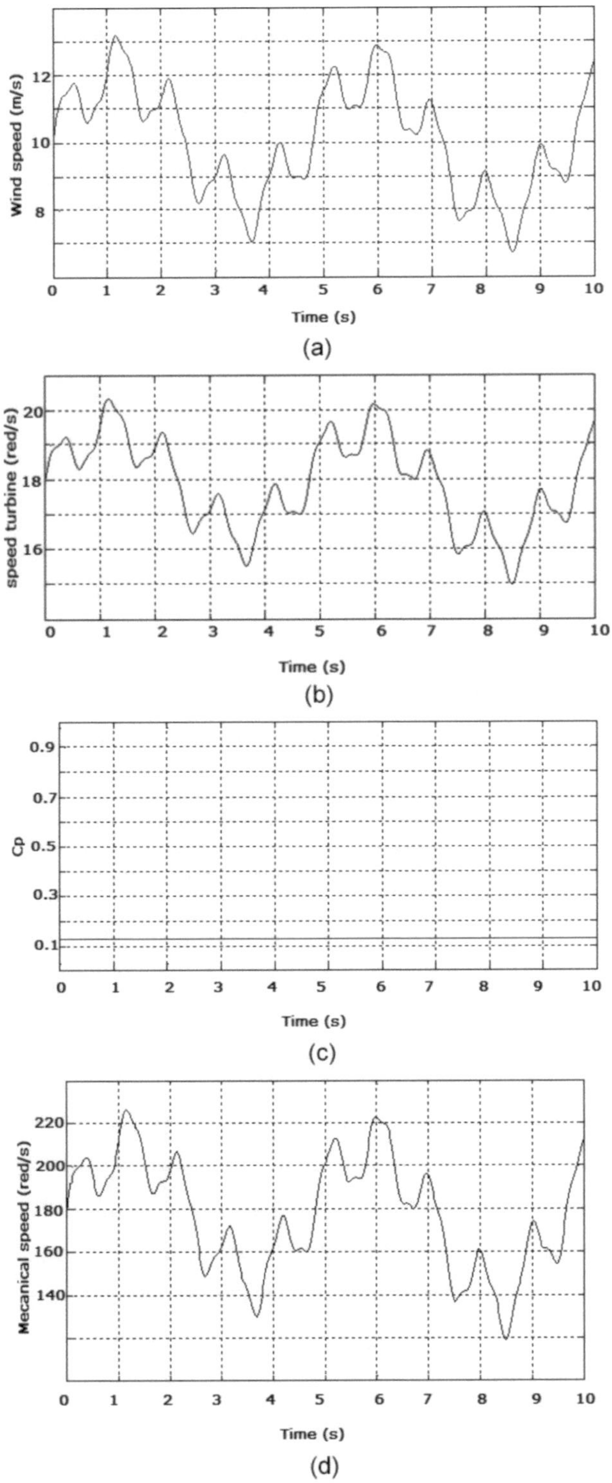

Figure 9.
The results of the wind turbine simulation. (a) The wind speed as a function of time. (b) The speed of the turbine as a function of time. (c) Power factor. (d) The electrical speed of the GSAP as a function of time.

The PWM rectifier was studied by simulation under MATLAB/SIMULINK according to the diagram of **Figure 10**, the current was controlled in the reference frame abc by hysteresis regulators, the reference of the voltage at the output of the

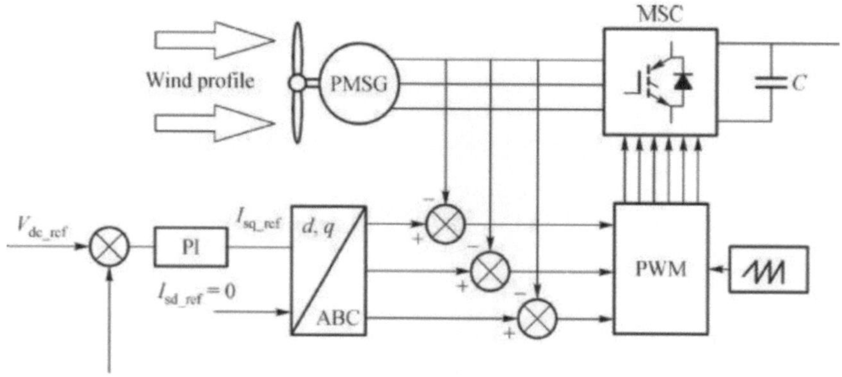

Figure 10.
Structure of control the wind system.

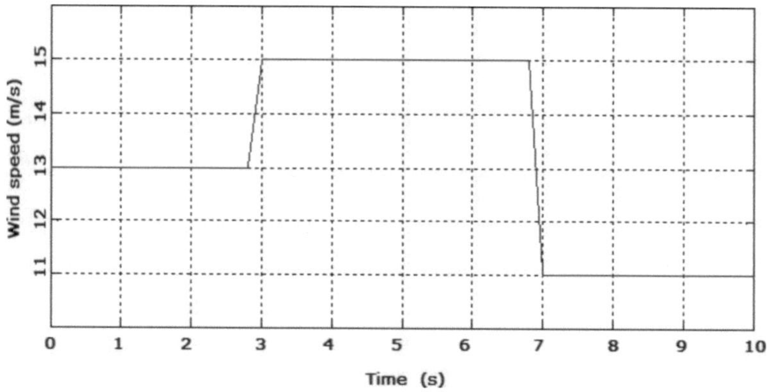

Figure 11.
Wind speed (m/s).

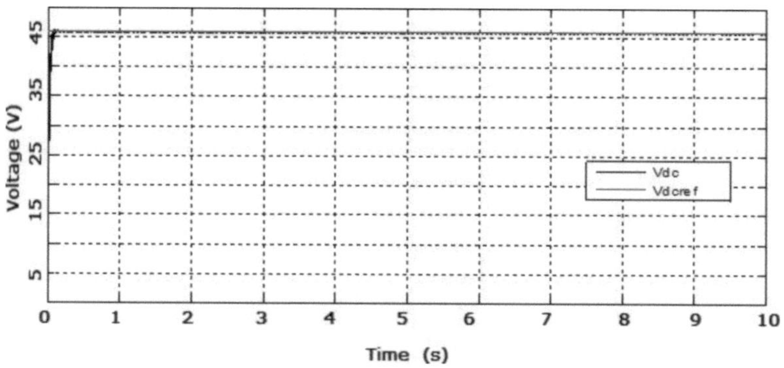

Figure 12.
Voltage rectified.

rectifier is taken equal to 45.8 V, We simulated the model of wind profile is given by **Figure 11**. **Figure 12** shows the DC voltage at the output of the rectifier with its proposed reference, it is observed that the DC bus voltage remains on average equal to 45.8 V. **Figure 13** shows the zoom of the rectified voltage.

According to the simulation results obtained, we conclude that the response of the voltage at the rectifier output (rectified) to a speed variation is relatively fast and does not exceed 2% of the reference value, during disturbances.

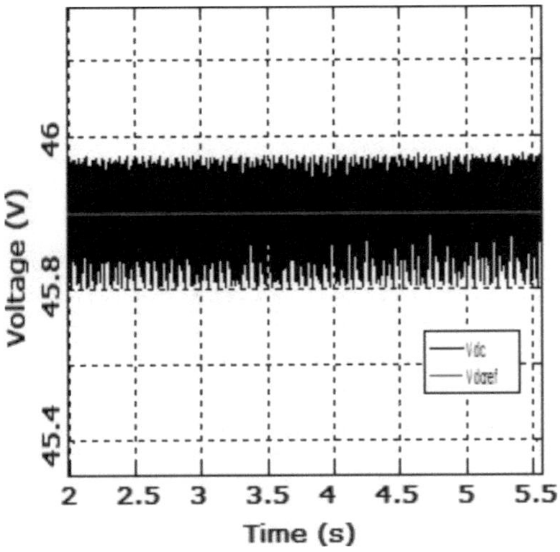

Figure 13.
Zoom of voltage rectified.

In spite of the increase or the decrease of the wind speed, the shape of the voltage of the continuous bus is established with 45.8 V with a time of response which depends on the control of the rectifier, of the order of 0.5 s in the example treated.

5. Conclusion

This chapter has established a global model of the wind energy conversion chain which consists of a GSAP permanent magnet synchronous machine associated with a wind turbine. The turbine used is the vertical axis Savonius type. They presented the simulation results of the complete wind energy production chain. The end of this chapter presents the application of the control of this wind energy production system. This control shows that the response of the voltage at the rectifier output (rectified) to a speed variation is relatively fast. In spite of the increase or decrease in wind speed, the speed of the continuous voltage is established with a response time which depends on the control of the rectifier.

Acknowledgements

Authors acknowledge the financial support of the General Direction of Scientific Research and Technological Development, DGRSDT.

Author details

Harrouz Abdelkader*, Fadila Tahiri, Boussaid Brahim and Fatiha Bekraoui
Department of Hydrocarbons and Renewable Energies, Faculty of Sciences and Technology, LDDI Laboratory, University of Ahmed Draïa, Adrar, Algeria

*Address all correspondence to: harrouz.onml@gmail.com

References

[1] Harrouz A, Benatialah A, Harrouz O. Direct power control of a PMSG dedicated to standalone wind energy systems. In: Journal IEEE Xplore, Eighth International Conference and Exhibition on Ecological Vehicles and Renewable Energy, March 27–30, 2013, Monaco, France. 2013

[2] Premalatha M, Abbasi T, Abbasi SA. Wind energy: Increasing deployment rising environmental concerns. Renewable and Sustainable Energy Reviews. Mar. 2014;**31**:270-288

[3] Mahersi E, Khedher A, Mimouni F. The wind energy conversion system using PMSG controlled by vector control and SMC strategies. International Journal of Renewable Energy Research. 2013;**3**(1)

[4] Harrouz A, Benatiallah A, Harrouz O. Modeling of small wind energy based of PMSG in south of Algeria. In: Journal "IEEE Explore" of 2nd International Symposium on Environment Friendly Energies and Applications (EFEA 2012). 2012. pp. 191-195. DOI: 10.1109/EFEA.2012.6294042

[5] Alnasir Z, Kazerani M. An analytical literature review of stand-alone wind energy conversion systems from generator viewpoint. Renewable and Sustainable Energy Reviews. Dec. 2013; **28**:597-615

[6] Tran D. Conception Optimale Intégrée d'une chaîne éolienne « passive ». In: Analyse de Robustesse, Validation Expérimentale. 2010. pp. 9-10

[7] Jedli H, Hidouri 2Noureddine. A power drive scheme for an isolated pitched wind turbine water pumping system based on DC machine. Journal of Advances in Mechanical Engineering and its Applications. March 2012;**1**:1

[8] Kendouli F, Nabti K, Abed K, Benalla H. Modélisation, simulation et contrôle d'une turbine éolienne à vitesse variable basée sur la géneratrice asynchrone à double alimentation. Revue of Renewable Energies. 2011; **14**(1):109-120

[9] Urtasun A, Idoia PS, Martín S, López J, Marroyo L. Modeling of small wind turbines based on PMSG with diode bridge for sensorless maximum power tracking. Renewable Energy Journal. July 2013;**55**:138-149

[10] Chinchilla M, Arnaltes S, Burgos JC. Control of permanent magnet generators applied to variable-speed wind-energy systems connected to the grid. IEEE Transactions on Energy Conversion. 2006;**21**(1):130-135

[11] Mahdi AJ, Tang WH, Jiang L, Wu QH. A comparative study on variable-speed operations of a wind generation system using vector control. In: Proceedings of ICREPQ'10, Granada (Spain). 2010

[12] Tawfiq K et al. Wind energy conversion system topologies and converters: Comparative review. Energy Procedia. 2019;**162**:38-47

[13] Li P, Hu W, Hu R, Huang Q, Yao J, Chen Z. Strategy for wind power plant contribution to frequency control under variable wind speed. Renewable Energy. 2019;**130**:1226-1236

[14] Zemamou M et al. Review of savonius wind turbine design and performance. Energy Procedia. 2017; **141**:383-388

Chapter 7

Implement Using KY Converter for Hybrid Renewable Energy Applications: Design, Analysis, and Implementation

Pushpavalli Murugan and Jothi Swaroopan Nesa Mony

Abstract

This chapter mainly focuses on meeting the energy demand and methodologies of renewable energy. Nowadays, researchers are mainly focusing on renewable energy from the sun, wind, biomass, etc. due to energy crises and the lack of non-renewable energy. The potential for solar energy is high and this demand can best be met with hybrid systems, which can provide an uninterruptible power supply. This chapter looks at the performance metrics of hybrid energy as well as the methodologies and various control techniques connected with power management. The chapter also defines the photovoltaic (PV)-based, novel, dual KY boost converter. Dual PV sources act as input for the dual KY boost converter to generate as much energy as possible from the dual PV system, using the inverter module to produce single-phase alternating current output. A dual KY boost converter can provide higher maximum power, a faster response, and smaller voltage ripple. KY boost converters are designed to generate stable output values according to various conditions because of various control techniques and the maximum power point tracking control algorithm.

Keywords: hybrid energy, grid system, control techniques, dual KY converter

1. Introduction

In India 53% of energy is produced from thermal power stations. The major problems with thermal energy are lack of coal, ash production, fuel emission, high running costs, and disruption to the ecology. This chapter mainly focuses on renewable energy sources like photovoltaic (PV), wind, and stable conversion using a hybrid system. In this introduction, the focus is on wind energy-related challenges because wind energy is not a linear energy. Wind is always fluctuating in nature so an appropriate controller is always required for optimal power. Here, an appropriate controller for an induction generator to achieve steady-state output will be described. An induction generator acts as a drive and a neural network controller is introduced with an ant colony optimization method. The controller is intended to drive a turbine to extricate extreme power from the wind. This strategy is dependable and reliable [1].

A wind energy system requires constant power at the conversion stage, and to avoid this variable, a frequency transformer technique is adopted. This methodology proposed in this chapter is a standalone wind system. The suggested strategy is modest, and has basic control with no harmonics [2]. The system consists of solar power, wind power, a diesel engine, and an intelligent power controller. To attain an active and balanced response for active power control, the controller consists of a radial basis function network and an improved Elman neural network for maximum power point tracking (MPPT). The pitch angle of the wind turbine and the PV system uses a radial basis function network, where the output signal is used to control the DC/Elman neural network DC boost converters to achieve MPPT [3]. Different types of permanent synchronous generator with various power capacity wind farms are connected to the common grid. This setup has three collected models of variable wind speed fitted with permanent magnet synchronous generator (PMSG) wind turbines for dynamic investigation [4].

This chapter mainly targets a suitable converter for a renewable energy system. It discusses the KY converter, which is suitable for hybridizing the energy. The KY converter gets its name from a paper titled "KY converter and its derivatives" by K.I. Hwu and Y.T. Yau (KY converter). A dual KY boost converter is presented, which is a KY converter combined with a normal synchronously rectified boost converter. The input and output inductor currents are continuous, with a higher voltage conversion ratio suitable for low-ripple applications. By using soft switching with the surge current suppressed the device can be used in high-power applications. The KY converter has better performance than a boost converter [5] and provides a fast response to reach the maximum power point compared to a boost converter. A novel voltage-bucking/boosting converter is called a KY buck-boost converter. It has fast transient responses, synchronous rectification, load regulations, and low-output voltage ripples. Cuk, Sepic, Zeta, and Luo converters could also be considered but they possess right half-plane zeroes, thereby causing system instability and slow load transient responses [6]. The operation of KY converters is also given, along with experimental results to verify ideal line and load regulations, and low output voltage ripples due to non-pulsating output currents.

Most solar panel energy efficiency ratings range from 15 to 20% [7]; therefore, various MPPT algorithms have been introduced. Numerous controllers are utilized to find MPPT and the efficiency is increased [8]. Power from any renewable energy source can be delivered to the grid individually or a balanced MPPT feature can be realized for renewable energy, power converters, and control algorithms along with purposeful energy resources for efficient operation of the microgrid. The proposed microgrid, based on hybrid energy resources, operates in autonomous mode and has an open architecture platform for testing multiple different control configurations. AC/DC, DC/AC, and DC/DC converters are integrated with the distributed energy system due to different types of output voltages [9]. Several algorithms that are used to operate DC/DC converters around the MPPT are reported in the literature. Comparing all algorithms, a fuzzy logic controller coupled to other controllers works well under partial shading conditions. There is also a newly designed fuzzy logic controller coupled to a Hopfield neural network maximum tracking technique [10]. The MPPT technique is very important because it increases the energy efficiency of a renewable energy system.

Another major challenge is hybrid renewable energy. The configuration of the hybrid renewable energy system and interfacing power converters for connecting the energy sources to the AC bus is extensively discussed. An outline of the control process in a hybrid renewable energy system and the application of the relevant control methods for system stabilization, effectively inducing real power and proper load-sharing methods, are available. Different approaches for hybrid renewable energy system design and control methods for power converters in the recent research literature are also briefly discussed. The AC bus-linked hybrid renewable energy system configuration reduces the number of power conversion stages and losses in power transferred to the load/utility. The master/slave control with the droop concept does not require a communication link and provides good load sharing, such as the flexibility, expandability, and modularity of the hybrid renewable energy system. Both single-master and multimaster approaches are used in the inverter control strategies [11]. In addition, a multi accumulated model utilizing comparable wind speed display has excellent similarity to the total model in transient and steady states and it also expends less time in reproduction when contrasted with the entire model [12]. Finally that the hybrid control system challenges not so distant future with its very own potential which clears another time with powerful and proficient energy balance. Globally, renewable energy has a green future and will undoubtedly become a credible alternative to fossil fuels [13, 14].

In conclusion, hybrid energy devices combining a variable velocity wind turbine and PV array-generating device could deliver a non-stop power to weight ratio to produce a viable hybrid controller. The hybrid controller manages the strength flow between the system's additives such that the cost is minimized and load necessities might be met throughout the year [15]. Hybrid renewable energy's main challenges are energy storage capacity and energy management. To utilize 100% renewable energy, the design requires intelligent charge control, battery state of charge, and estimation of the impact of various working routines on battery life. Different architectures of energy sources and power management on peak load sharing are analyzed with a reduction of varying losses [16].

The next important analysis concerns a multi-input system. A novel multi-input inverter is used for the grid-connected hybrid solar/wind power system to simplify the power system and reduce the cost. The multi-input inverter is designed with a buck/buck–boost fused to a multi-input DC/DC converter and a single-phase bridge-type DC/AC inverter [17]. A semi-isolated multi-input converter for a hybrid PV/wind power charger system that can simplify the power system, reduce cost, deliver continuous power, and overcome high-voltage-transfer ratio problems is proposed [18].

Various literatures on the use of hybrids with renewable energy systems for uninterrupted power supply to the load have been analyzed. Another requirement is the MPPT algorithm along with the multi-input converter, which is used to reduce the volume of the system. Hybrid renewable energy systems in combination with existing setups are presented to investigate the possibility of specialized aggressiveness [15]. In this chapter, the operational principle of the proposed semi-isolated multi-input converter is explained: a novel, executed multi-input KY boost converter for hybrid systems associated with the grid. The control parameters are reliant on the wind, sun, storage, and grid conditions and checked under dynamic conditions. DC bus voltages are used to deal with the contribution of various working mode controls amid different working conditions [19, 20].

2. Hybrid energy system methods

2.1 Types of integration system

There are numerous approaches to coordinate distinctive alternative energy to form the integration of hybrid systems. For the most part, strategies can be arranged into three classes: DC coupled, AC integrated, and hybrid integrated. The AC-integrated system can be grouped into power-frequency AC-integrated and high-frequency AC-integrated systems. The strategies of a DC-coordinated setup are shown in **Figure 1**.

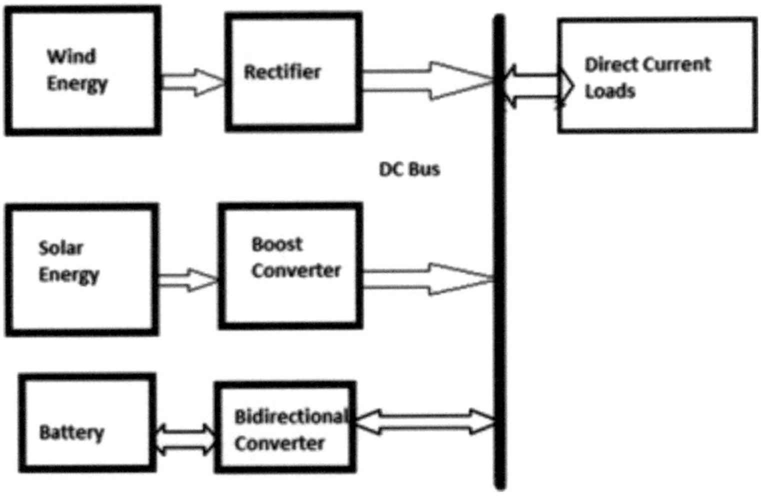

Figure 1.
Block diagram of a DC-coupled integration system.

2.1.1 DC-coupled integration system

The DC sources may be associated with the DC bus directly if they are suitable. If there are any DC loads, they can be clearly related to the DC bus, or through the DC/DC converters, to fit DC voltages to DC loads [21].

2.1.2 AC-coupled integration system

AC coupling can be split into two further categories: power frequency AC-coordinated and high-frequency AC-coordinated systems. An AC-coordinated system is shown in **Figure 2**, where diverse energy sources are coordinated through their control switching interfacing circuits to a power recurrence AC bus. Inductors may likewise be required between the control switching circuits and the AC bus to accomplish the desired control through the board.

2.1.3 Hybrid-coupled integration system

Rather than interfacing all the distributed generation sources to only one DC or AC bus, as already discussed, the distinctive distributed generation sources can be associated with the DC or AC bus of the mixture system. **Figure 3** demonstrates a hybrid-coupled coordinated system, where distributed generation assets are associated with the DC bus as well as the AC bus. In this arrangement, energy sources can be coordinated specifically without any additional circuits. Therefore, the framework can have higher energy effectiveness and diminished cost. Meanwhile, control and power flow management may become increasingly difficult compared to DC and AC bus-coordinated systems [22].

2.2 System arrangement

The system comprises renewable energy and alternative energy. The renewable energies are wind, solar, and biomass. Alternative energies are a diesel generator and a storage system. In a storage system, a battery, fuel cells, supercapacitors, and aqua electrolysis are used [23]. The level of dependability of a PV and wind-coordinated framework to suit a specific load can be given by a mix of

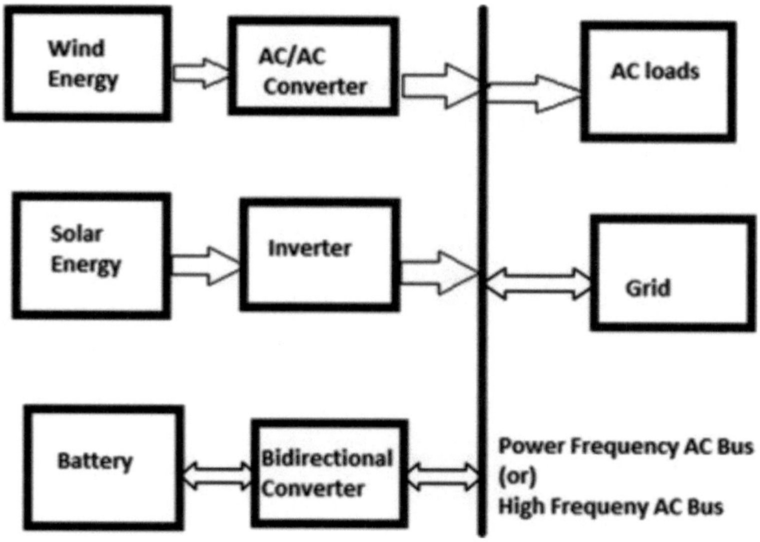

Figure 2.
Block diagram of an AC-coupled integration system.

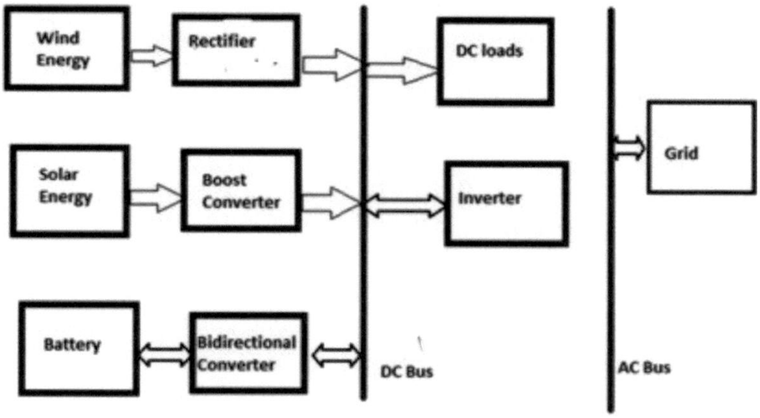

Figure 3.
Block diagram of a hybrid-coupled integration system.

appropriately estimated wind energy, PV, battery units, and alternative sources. In atmospheres, alternative sources are expected to be highly dependable also, keep away from gross over-plan of the sunlight based and wind framework. In other words, an assistant energy provider can be viewed as an optional technique for proposing an alternative energy source fused into the framework compared to expanding equipment sizes unnecessarily for those periods when energy generation is reduced [24]. The PMSG-based wind turbine works at the greatest power point following the MPPT mode to extract power from the wind, although wind power may need to be reduced under specific conditions. Wind reduction can be accomplished by utilizing the turbine's capacity and pitch control systems [25].

The execution of crystalline silicon PV modules limits the physical elements of the PV cell material, the temperature of PV array cells and the sun situated irradiance revealed on the sun's situated cells. One solved proper model for the most outrageous power yield of PV modules is used [4–7, 12, 19, 20, 22–24, 26–30].

Changes in wind speed and sunlight-based illumination are associated with the wind turbine generator, and PV can recreate a variety of intensities of AC and DC sources and test the MPPT control calculation [28].

Regarding the size of the battery bank to use for these sorts of uses requires examination of the battery's charge and discharge capabilities, including burden, yield, and in the case of sunlight-based or elective critical sources, the working temperature and capability of the charger and other parts of the structure. By and large, problems occur while charging the battery bank and its capability drops when the battery ages or is misused. At the point when the aggregate yield of solar and wind energy is more than the energy requested, the storage unit is charged. Two properties of the battery are associated with the hybrid system's execution, i.e. the state of charge and the charge voltage [29].

The hybrid system grid plays a vital role and can work in two modes. Grid Tied mode the power converters to produce constant voltage to a DC bus and exchange of power can be accomplished using DC and AC buses, while the grid working as an autonomous mode storage unit plays a vital role in balancing voltage and power.

2.3 Expert control

2.3.1 Intelligent controller

The control framework for hybrid energy system designs should limit fuel utilization by expanding energy from inexhaustible sources. Nonetheless, there are control changes by which the sustainable power source can change, which causes unsettling influences that can influence the nature of the energy conveyed to the heap. To deal with the stream of energy effectively, with fewer harmonics, a controlling strategy should be developed.

The energy sources are modeled from PV, wind, and battery. To coordinate all the converters used in the hybrid system, intelligent controllers are essential. A proportional-integral controller, fuzzy logic controller, neural network, and genetic algorithms are used as an intelligent controller [7].

2.3.2 Control methods

- The hybrid system has three types of controller. The controller used in PV and wind is said to be a generator side converter to obtain maximum power point tracking from solar and wind.

- A grid side converter is used to sustain constant voltage in a DC bus and frequency in an AC bus, to control the exchange of reactive power in the AC bus.

- A storage side converter maintains constant DC voltage and state of charge of the battery.

These converters will coordinate the entire hybrid system.

2.3.3 Power management

Power management is categorized as two types: rule based and optimization based. Rule based mainly depends on real-time applications. Depending on the grid requirements, power flow will be controlled from source to grid. Optimization based depends on the cost of energy. Its main focus is to minimize cost.

As the arrangement of hybrid renewable energy and alternative energy systems as a microgrid expands, so does the necessity for continuous energy for such systems. Powerful correspondence between the separate energies of the microgrid is imperative and merits further consideration. In addition, efficient methodologies and institutionalization are required for the production and safe organization of such systems.

2.3.4 Grid-connected mode

At the point where the hybrid grid works in islanding mode, the boost converter and the consecutive AC/DC/AC converter of the doubly-fed induction generator may work in on-MPPT or off-MPPT mode depending on framework control equality and criticality goals. The essential converter is a voltage source that gives a consistent voltage that repeats for the AC framework and works either in inverter or converter mode for smooth power exchange between the AC and DC link. The battery converter works either in charging or discharging mode and is reliant on the power balance in the framework. DC interface voltage is maintained by either the battery or the boost converter and is reliant on structure working conditions. Forces under different load and supply conditions should be adjusted.

When the hybrid system works in grid-connected mode, the goal of the boost converter is to pursue the MPPT of the PV display by coordinating its terminal voltage. The sequential AC/DC/AC converter of the doubly-fed induction generator is controlled to guide the rotor side current to achieve MPPT and to synchronize with the AC framework. The energy overflow of the hybrid grid framework can be sent to the utility framework. The activity of the battery for energy stockpiling is less imperative since the power is adjusted by the utility framework. For this situation, the fundamental purpose of the battery is to eliminate customary control trade between the DC and AC link. The DC/DC converter of the battery can be controlled as an energy pad using this strategy. The principal of converter is intended to work bidirectionally to fuse corresponding normal for wind what's more, sun-powered sources. The goals of the primary converter are to maintain a steady DC-connected voltage for the variable DC stack and to synchronize with the AC connection and utility framework.

3. Design of a dual KY boost converter

3.1 Working of a dual KY boost converter

Figure 4 demonstrates the proposed dual KY boost converter built using the KY converter joined with the boost converter. The dual KY converter is made using four switches M1, M2, M3, and M4, one diode D, one output inductor Lo, two input inductors Li1, Li2, a support capacitor Cs, and one output capacitor Co. The depository capacitor Cs is a support between the KY converter and the conventional synchronously rectified boost converter. The multi-input DC voltage is 12 V and 110 V are obtained at the yield. Because of the conduction status of the switches the dual KY converter has two modes of operation. **Figure 5** shows mode 1 operation of the dual KY boost converter. M1 and M4 are turned off and M2 and M3 are turned on. The diode D is forward biased and begins to conduct, and Cs is released. In this manner, the voltage across Li1 is V1, making Li1 charged, while the voltage across Lo is Vo minus VCs, subsequently making Lo demagnetized. Additionally, the current flowing through Co is equivalent to ILo minus the current moving through RL.

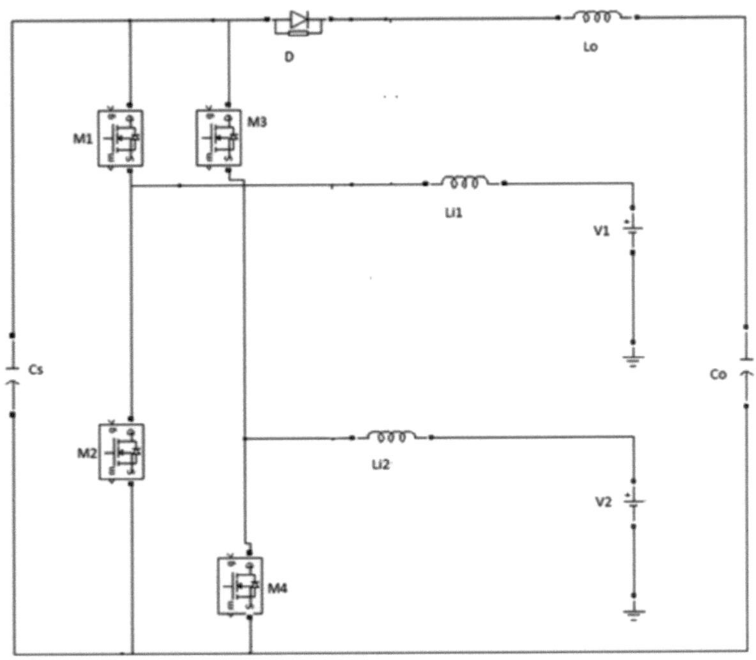

Figure 4.
Dual KY boost converter connected to DC input.

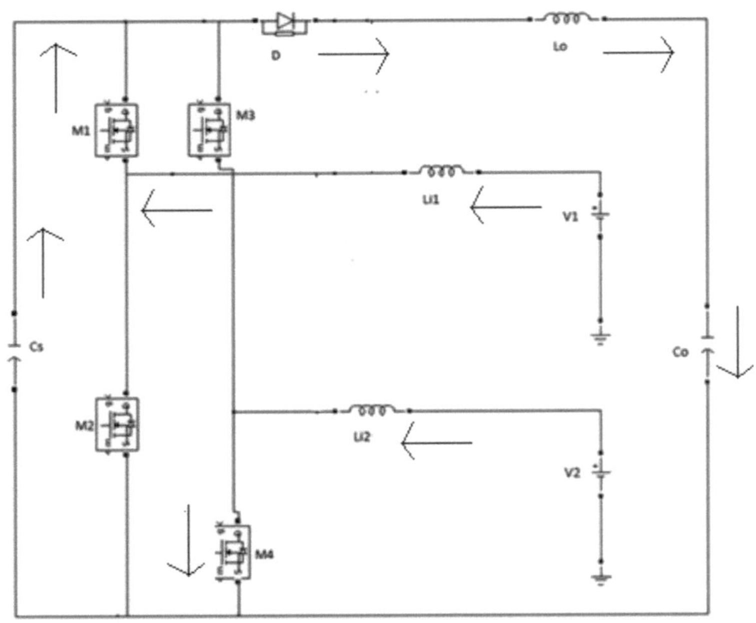

Figure 5.
Mode 1 operation of the dual KY boost converter.

Figure 6 shows mode 2 operation of a dual KY boost converter. M2 and M3 are turned on and M1 and M4 are turned off. Consequently, the voltage across Li1 is VCs subtracted from voltage source V1, subsequently making Li1 demagnetized, while the voltage across Lo is Vo subtracted from 2VCs making Lo polarized. Additionally, the current flowing through Co is equivalent to ILo minus the current moving through RL, while the current moving through Cs is equivalent to the total of ILi1 and −ILo.

The input voltage V1 = V2 = 12 V and the input power of 12 W is made up from each source of 6 W. **Figure 8** shows the output power of a dual KY boost converter of 22 W with output voltage of 110 V. The results are shown in **Figures** 7 and **8**.

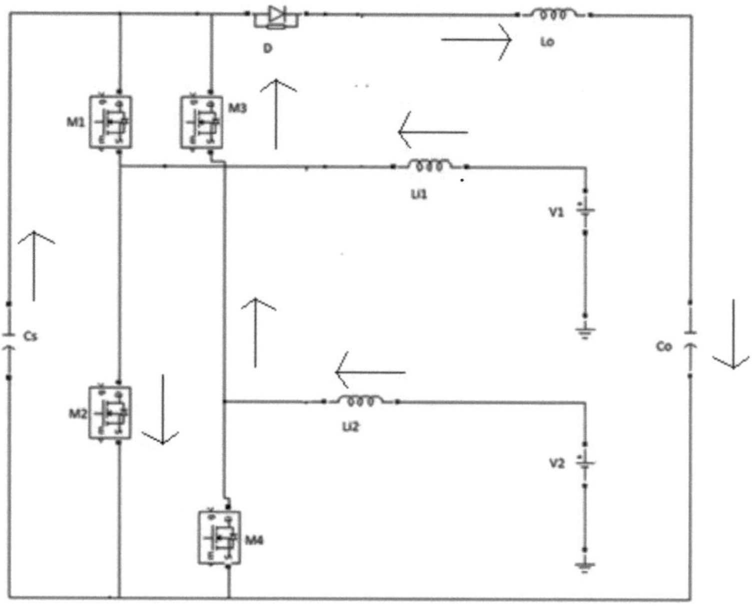

Figure 6.
Mode 2 operation of a dual KY boost converter.

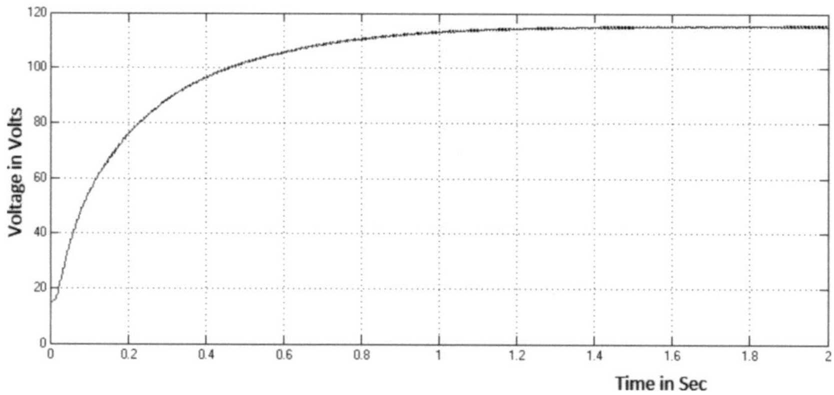

Figure 7.
Dual KY boost converter output voltage is 110 V.

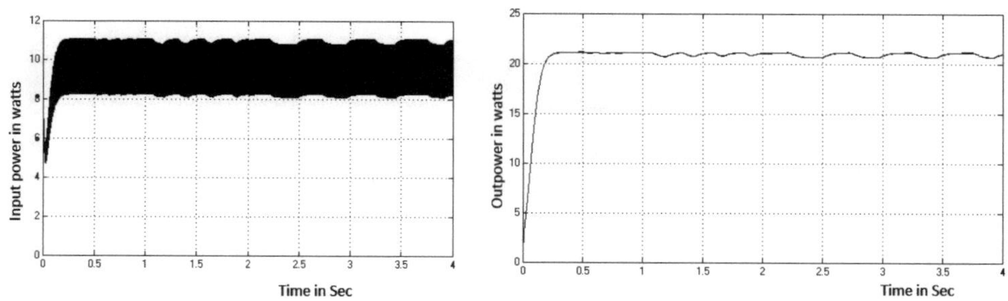

Figure 8.
Total input power = 12 W and dual KY converter output power = 22 W.

3.2 Disadvantages of the existing method

Regular converters for interfacing numerous sustainable sources utilize a typical DC connection or normal AC recurrence interface, which has disadvantages like the necessity for increased bundling, and expanded switches over the ordinary structure becoming delineated [31]. The multiport converter has numerous points of interest compared to the traditional structure regarding the number of energy gadgets and transformation steps that can enhance system cost. Certain non-segregated voltage-bucking/boosting converters make the system cost effective. The interleaved converter can likewise be utilized. It comprises two single-stage support converters in parallel.

3.3 Advantages of the proposed method

This dual KY support converter topology highlights higher proficiency with less powerful hardware gadgets and fewer power change processes. The proposed dual KY support converter has the benefits of basic topology and the least number of energy switches. The dual KY support converter has persistent data and yields inductor streams and high-voltage proportions. The dual KY support converter structure is promising from the perspectives of minimal effort, concentrated control, and small size [27].

4. Dual KY boost converter connected to inverter

The dual KY converter is connected to a single-phase inverter. Analysis is done for a PV array input of 18 V given to V1 and V2 and the KY converter voltage is 230 V with a current of 3.5 A. The dual KY boost converter output is connected to a single-phase inverter. The single-phase inverter output voltage is 230 V and the total harmonic distortion (THD) value is nearly 8.22%.

4.1 Dual KY boost converter connected to PV array input

Table 1 listed the design value of PV array.

Parameters	Values
Isc	5.96 A
Voc	6.2 V
Series resistance	0.18 ohm
Parallel resistance	2 ohm
No. of modules connected in series	4
No. of modules connected in parallel	4
Irradiance	1000 W/m^2
Temperature	25°C

Table 1.
Design value of PV array.

4.2 Simulation results of dual KY boost converter connected to inverter

The above design values of PV arrays are connected as input and the waveforms are obtained. The analysis is done for a PV array input of 18 V shown in **Figures 9** and **10** and an output of 230 V obtained across the load resistance shown in **Figure 11**. The component determinations are given as follows: (1) evaluated PV array input voltage is 18 V; (2) DC voltage is 110 V; (3) exchanging recurrence is 2 kHz; (4) estimation of Li1 and Li2 is 400 μH and the estimation of Lo is the same as that of Li; (5) one 1 μF capacitor is decided for Cs; (6) one 470 μF capacitor is decided for Co; and (7) parameters of the PI controller, kp and ki, are set to 1 and 10 individually. These outputs are replicated using the simulation shown in **Figures 12** and **13**.

Figure 14 shows the perturbation and observation algorithm used to generate the duty ratio for switches. The simulation also shows a dual KY boost converter connected to battery terminals and the state of charge is checked. The dual KY boost converter output is connected to a single-phase inverter. The single-phase inverter output voltage is 230 V and the THD value is nearly 8.22%.

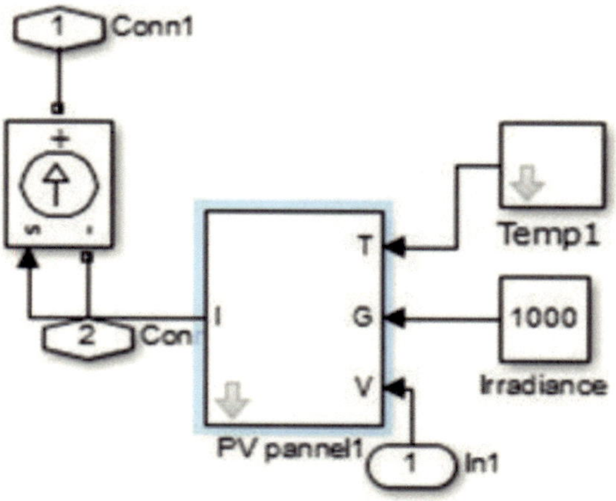

Figure 9.
Dual KY boost converter connected to PV array input.

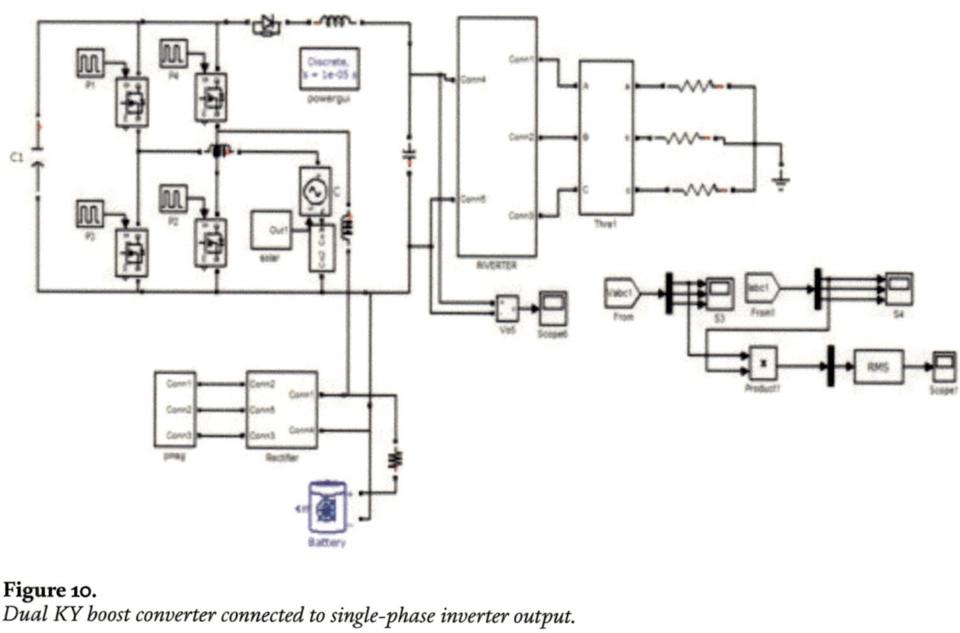

Figure 10.
Dual KY boost converter connected to single-phase inverter output.

The corresponding waveforms are shown in **Figures 15–18**. **Table 2** shows the output values of KY boost converter. **Table 3** shows the comparison of KY boost and boost converter.

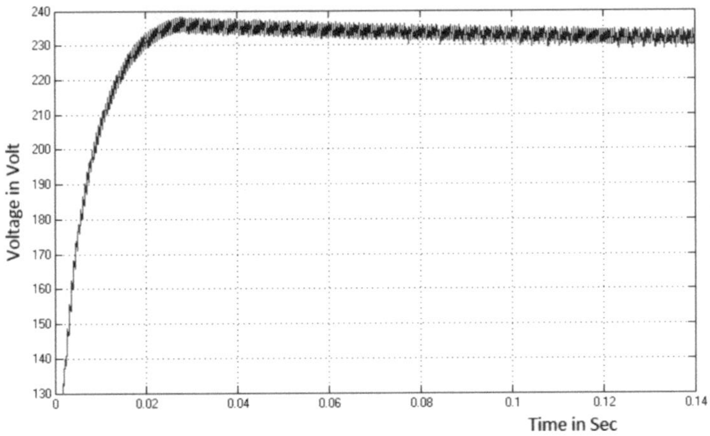

Figure 11.
Dual KY boost converter DC output voltage of 230 V.

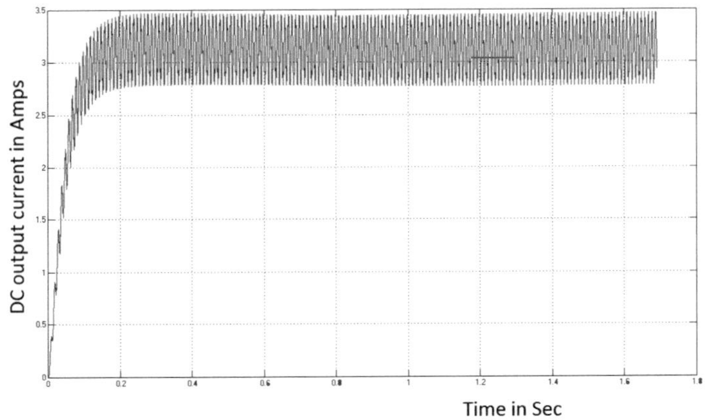

Figure 12.
Dual KY boost converter DC output current of 3.5 A.

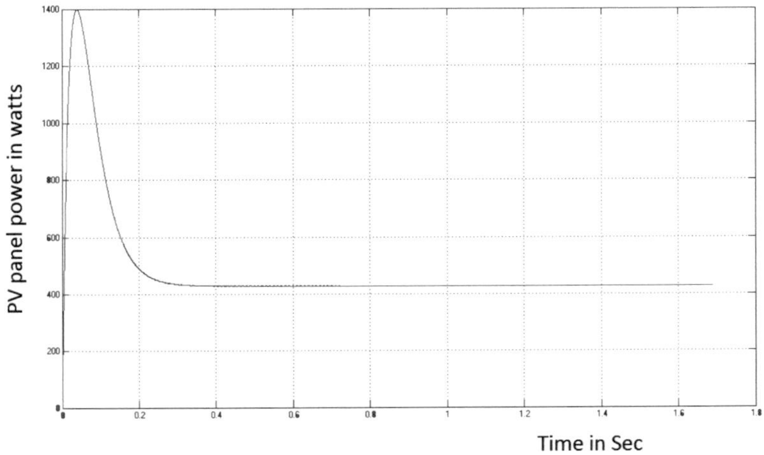

Figure 13.
Ppv = 450 W.

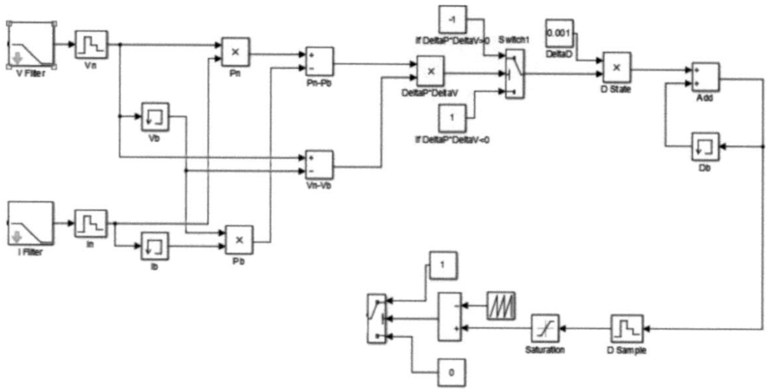

Figure 14.
Perturbation and observation algorithm for generating pulses.

Parameters	Values
Isc	26.3 A
Voc	19.5 V
PV panel voltage	18 V
PV panel current	25A
PV panel power	450 W
KY boost converter voltage	230 V
KY boost converter current	3.5 A
Inverter AC output voltage	230 V
Inverter AC output current	5 A
Power factor	0.93

Table 2.
Output of KY boost converter.

Converter	Vin	Vo	THD (%)
KY boost converter	18 V	230 V	6.58
Boost converter	18 V	110 V	10

Table 3.
Comparison of output voltage and THD.

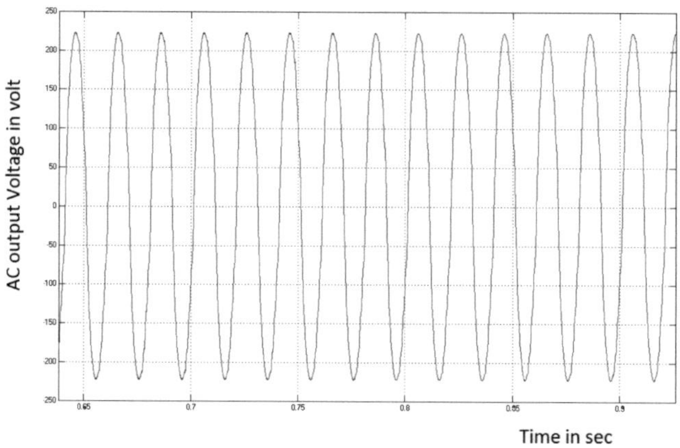

Figure 15.
Inverter AC output voltage is 230 V.

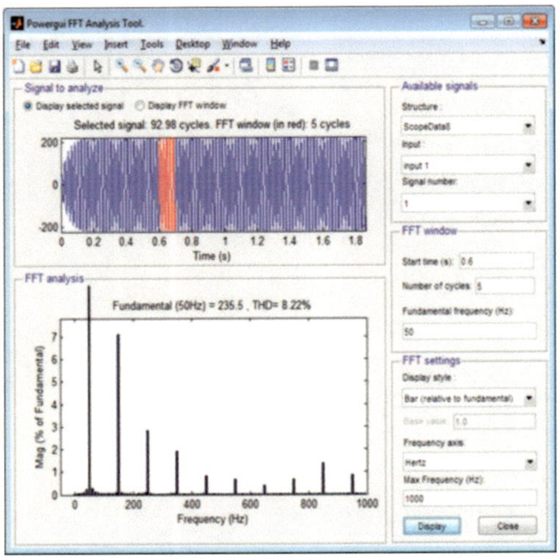

Figure 16.
THD = 8.22% for single-phase inverter output voltage.

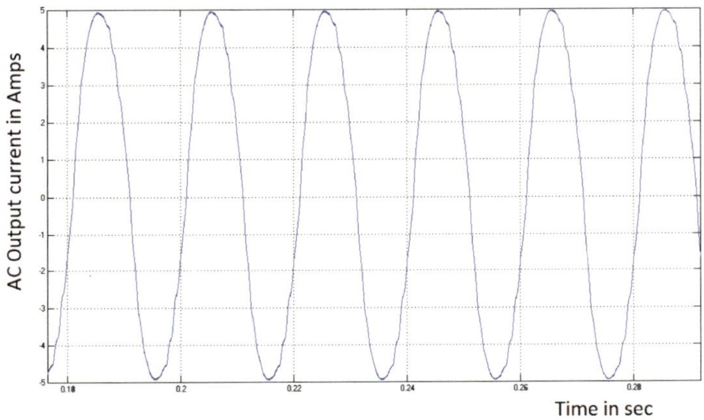

Figure 17.
Single-phase inverter AC output current is 5 A.

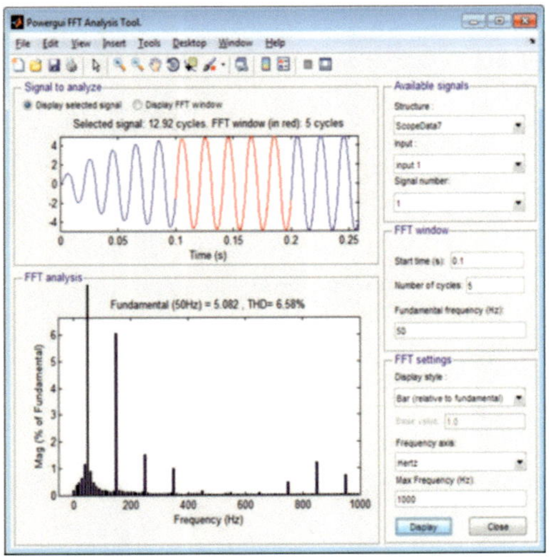

Figure 18.
THD = 6.58% for single-phase inverter output current.

DOI: http://dx.doi.org/10.5772/intechopen.90755

5. Conclusion

Because of fast improvements in the field of sustainable power sources and expanding costs of ordinary items like oil and gas, the use of sustainable power sources has turned out to be exceptional in hybrid systems. Solar PV and wind systems have been promoted around the globe on a comparatively large scale. This chapter proposed a hybrid energy system consisting of a wind turbine and a PV source to supply continuous power to the load. A controller is used to track the maximum power from renewable sources. Simple and economic control with a KY converter, which is a DC/DC converter, is used for maximum power extraction from the wind turbine and PV array. Power from the dual PV array can be transported to the utility grid. At the same time, the MPPT attribute can be acknowledged for the dual PV system. The substantial range of PV array voltage, current, and temperature variations was caused by various insolations. This solitary dual generation system can adequately separate extremes of power from the two PV sun-oriented energy sources. Boosting the DC voltage (18–230 V) to an adequate level utilizing the dual KY boost converter and obtaining pure AC voltage (230 V) from the inverter are key to understanding the above targets. A model of a hybrid dual solar system arrangement utilized with a control strategy has therefore been created. The KY converter will offer low THD (6.58%), high power factor under nonlinear stacking conditions, and pronounced unique reactions under transient stacking conditions.

Author details

Pushpavalli Murugan[1*] and Jothi Swaroopan Nesa Mony[2]

1 Research Scholar/EEE, Sathyabama Institute of Science and Technology, Chennai, India

2 Professor/EEE, RMK Engineering College, Chennai, India

*Address all correspondence to: pushpavalli.eee@sathyabama.ac.in

References

[1] Hong C-M, Cheng F-S, Chen C-H. Optimal control for variable-speed wind generation systems using general regression neural network. International Journal of Electrical Power & Energy Systems. 2014;**60**:14-23

[2] Bakhsh FI, Khatod DK. A new synchronous generator based wind energy conversion system feeding an isolated load through variable frequency transformer. Renewable Energy. 2016;**86**:106-116

[3] Lin WM, Hong CM, Chen CH. Neural-network-based MPPT control of a stand-alone hybrid power generation system. IEEE Transactions on Power Electronics. 2011;**26**(12):3571-3581

[4] Badr MA, Atallah AM, Bayoumi MA. Comparison between aggregation techniques for PMSG wind farm. Energy Procedia. 2015;**74**:1162-1173

[5] Hwu KI, Yau YT. Two types of KY buck–boost converters. IEEE Transactions on Industrial Electronics. 2009;**56**(8):2970-2980. DOI: 10.1109/TIE.2009.2023100

[6] Hwu KI, Yau YT. Topology exchange between KY converter and its derivative. In: 2008 Twenty-Third Annual IEEE Applied Power Electronics Conference and Exposition. Austin, TX, USA: IEEE; 2008. pp. 1675-1678

[7] Deshmukh MK, Deshmukh SS. Modeling of hybrid renewable energy systems. Renewable and Sustainable Energy Reviews. 2008;**12**(1):235-249

[8] Ponkarthik N, Kalidasa Murugavel K. Performance enhancement of solar photovoltaic system using novel maximum power point tracking. International Journal of Electrical Power & Energy Systems. 2014;**60**:1-5

[9] Anbarasi Jebaselvi GD, Paramasivam S. Analysis on renewable energy systems. Renewable and Sustainable Energy Reviews. 2013;**28**:625-634

[10] Subiyanto S, Mohamed A, Hannan MA. Intelligent maximum power point tracking for PV system using Hopfield neural network optimized fuzzy logic controller. Energy and Buildings. 2012;**51**:29-38

[11] Senjyu T, Nakaji T, Uezato K, Funabashi T. A hybrid power system using alternative energy facilities in isolated island. IEEE Transactions on Energy Conversion. 2005;**20**(2):406-414

[12] Rolan A, Luna A, Vazquez G, Aguilar D, Azevedo G. Modeling of a variable speed wind turbine with a permanent magnet synchronous generator. In: 2009 IEEE International Symposium on Industrial Electronics. Seoul, South Korea: IEEE; 2009. pp. 734-739

[13] Ekren BY, Ekren O. Simulation based size optimization of a PV/wind hybrid energy conversion system with battery storage under various load and auxiliary energy conditions. Applied Energy. 2009;**86**(9):1387-1394

[14] Yang H, Lu L, Zhou W. A novel optimization sizing model for hybrid solar-wind power generation system. Solar Energy. 2007;**81**(1):76-84

[15] Nema P, Nema RK, Rangnekar S. A current and future state of art development of hybrid energy system using wind and PV-solar: A review. Renewable and Sustainable Energy Reviews. 2009;**13**(8):2096-2103

[16] Bocklisch T. Hybrid energy storage systems for renewable energy

applications. Energy Procedia. 2015;**73**:103-111

[17] Chen YM, Liu YC, Hung SC, Cheng CS. Multi-input inverter for grid-connected hybrid PV/wind power system. IEEE Transactions on Power Electronics. 2007;**22**(3):1070-1077

[18] Chen CW, Chen KH, Chen YM. A semi-isolated multi-input converter for hybrid PV/wind power charger system. In: 2014 International Power Electronics Conference (IPEC-Hiroshima 2014-ECCE ASIA). Hiroshima, Japan: IEEE. pp. 3592-3597

[19] Pushpavalli, M. Swaroopan NJ. Maximum power tracking for PV array system using fuzzy logic controller. In International Conference on Intelligent Computing and Applications. Springer: Singapore; 2019. pp. 337- 348

[20] Pushpavalli M. Harmonic reduction study in KY boost and single phase AC converter. International Journal of Applied Engineering Research (IJAER). 2015;**10**:3549-3558

[21] Xu L, Chen D. Control and operation of a DC microgrid with variable generation and energy storage. IEEE Transactions on Power Delivery. 2011;**26**(4):2513-2522

[22] Adhikari S, Li F. Coordinated V-f and P-Q control of solar photovoltaic generators with MPPT and battery storage in microgrids. IEEE Transactions on Smart Grid. 2014;**5**(3):1270-1281

[23] Nehrir MH, Wang C, Strunz K, Aki H, Ramakumar R, Bing J, et al. A review of hybrid renewable/alternative energy systems for electric power generation: Configurations, control, and applications. IEEE Transactions on Sustainable Energy. 2011;**2**(4):392-403

[24] Lau KY, Yousof MFM, Arshad SNM, Anwari M, Yatim AHM. Performance analysis of hybrid photovoltaic/diesel energy system under Malaysian conditions. Energy. 2010;**35**:3245-3255

[25] Jaramillo-Lopez F, Kenne G, Lamnabhi-Lagarrigue F. A novel online training neural network-based algorithm for wind speed estimation and adaptive control of PMSG wind turbine system for maximum power extraction. Renewable Energy. 2016;**86**:38-48

[26] Ramaprabha R, Balaji M, Mathur BL. Maximum power point tracking of partially shaded solar PV system using modified Fibonacci search method with fuzzy controller. Electrical Power and Energy Systems. 2012;**43**:754-765

[27] Pushpavalli M, Jothi Swaroopan NM. Performance analysis of hybrid photovoltaic/wind energy system using KY boost converter. International Journal of Power Electronics and Drive Systems (IJPEDS). 2019;**10**(1):443

[28] Murtaza A, Chiaberge M, De Giuseppe M, Boero D. A duty cycle optimization based hybrid maximum power point tracking technique for photovoltaic systems. Electrical Power and Energy Systems. 2014;**59**:141-154

[29] Dali M, Belhadj J, Roboam X. Hybrid solar-wind system with battery storage operating in grid-connected and standalone mode: Control and energy management—Experimental investigation. Energy. 2010;**35**(6):2587-2595

[30] Kim HS, Kim JH, Min BD, Yoo DW, Kim HJ. A highly efficient PV system using a series connection of DC–DC converter output with a photovoltaic panel. Renewable Energy. 2009;**34**(11):2432-2436

[31] Liu X, Wang P, Loh PC. A hybrid AC/DC microgrid and its coordination control. IEEE Transactions on Smart Grid. 2011;**2**(2):278-286

Chapter 8

Survey on Photo-Voltaic Powered Interleaved Converter System

M.L. Bharathi

Abstract

Renewable energy is the best solution to meet the growing demand for energy in the country. The solar energy is considered as the most promising energy by the researchers due to its abundant availability, eco-friendly nature, long lasting nature, wide range of application and above all it is a maintenance free system. The energy absorbed by the earth can satisfy 15000 times of today's total energy demand and its hundred times more than that our conventional energy like coal and other fossil fuels. Though, there are overwhelming advantages in solar energy, It has few drawbacks as well such as its low conversion ratio, inconsistent supply of energy due to variation in the sun light, less efficiency due to ripples in the converter, time dependent and, above all, high capitation cost. These aforementioned flaws have been addressed by the researchers in order to extract maximum energy and attain hundred percentage benefits of this heavenly resource. So, this chapter presents a comprehensive investigation based on photo voltaic (PV) system requirements with the following constraints such as system efficiency, system gain, dynamic response, switching losses are investigated. The overview exhibits and identifies the requirements of a best PV power generation system.

Keywords: solar, renewable energy, converters

1. A survey on soft switching technique

The efficiency of our conventional boost converters is mainly affected by the power loss during switching. The hard switching performed by the switches actually causes major power lose and hard-switching of the semiconductor switches, generates voltage and current stress during turn-on and turn-off process. In order to overcome the losses during the switching stress, plenty of methods have been derived. Among these methods, however, soft-switching is considered to be the most popular and most efficient method. The soft switching technique is introduced in early 90s and soft switching means smooth turn on and turn off of switches.

During soft switching, zero voltage switching (ZVS) and zero current switching (ZCS) is possible during turn-on and turn off of the power switches. Interleaved converters have showed a significance improvement in the efficiency when compared to the conventional boost converter which is evident from our previous research work. Some of the soft switching techniques adopted till now are summarized briefly in **Table 1**. This section discusses and reviews the performance of few soft switching topologies that were introduced in the recent past.

IntechOpen

From the above literature, it is evident that resonant soft switching technique with power conditioning system could be considered as the best due to its reduced switching losses and its ability to attain high efficiency with no additional switches for soft switching. The objective of high efficiency and better performance is achieved by connecting PV power conditioning system (PVPCS), MPPT using PV simulator and current balancing control technique. Most of the above-mentioned work deals with soft switching with an external circuit to achieve ZVS and ZCS condition of the switches for boost, buck and also for buck-boost. Since switches experience less voltage stress across them, the conduction loss is reduced to a great extent which obviously improves the efficiency.

Though, lossless switching have obtained improved efficiency care should be taken to reduce the converter size since extra external device are used for soft switching and also make the converter cost effective. Interleaved converters have filled this gap of over size of converters by avoiding external device rather, it designed and emphasized in such a way that two or more converters are put together to obtain improved efficiency and in turn voltage gain.

Cha et al. [1] proposed a solar powered high efficiency soft switching boost converter. Soft switching is achieved by adding an additional switch of 30 kHz switching frequency IGBT, a couple of diodes and single capacitor to the conventional boost converter. By adopting this method, the size of system is reduced.

Wu et al. [2] eliminated hard switching by adopting active clamping technique in 190 kHz switching frequency MOSFET. Where, ZCS cell is included with the conventional converter of 500 W, with 150 V DC input voltage. It consists of an auxiliary inductor and a clamping diode. The output of 350 V DC voltages is obtained.

Das et al. [3] achieved steep conversion ratio in the conventional converter of 200 W and 66 kHz switching frequency with the help of coupled inductor and eliminated switching losses by adding auxiliary circuit consists of a switch and passive elements. The converter can be operated in both buck and boost mode with input voltage of 30 V DC and 200 V DC voltage.

Park et al. [4] improved efficiency of our conventional converters of 700 W by eliminating the losses by soft switching. The converter input voltage is 230 V DC and output voltage is 380 V DC and has switching frequency of 30 kHz. A simple auxiliary resonant circuit (SARC) adopts ZCS during turn on and ZVS during turn off. SARC consist of two diodes, an inductor and a capacitor. It is easy control since both the switches are controlled by PWM signal. MPPT controller is applied to the circuit.

Lee et al. [5] used soft switching technique and interleaving technique in a buck converter to improve the step-down conversion ratio where the efficiency of the converter in high with an input of 200 V DC and output of 24 V DC with a switching frequency ranges from 65 kHz to 300 kHz. Here, the active switches connected in series and a capacitor is connected in the power path. The two active switches are connected in series with a coupled capacitor in the power path to yield higher conversion ratio. Since switches are connected in series the voltage and current stress across the switches are less and easy to control. High switching frequency, less voltage stress and reduced switching losses are achieved. Capacitive discharging is lowered. High conversion ratio and smaller output current ripple is obtained. An output voltage ripple of 250 mV is witnessed.

Amorndechaphon et al. [8] connected an active resonant snubber with a DC-DC boost converter in grid connected PV system in order to reduce the switching losses and switching stress. Circuit components for soft commutation are derived from equation of each operating stage. The proposed active resonant snubber experiences reduced switching losses as well as stress.

Lee et al. [9] handled losses in a conventional boost converter of 150–340 V DC voltage due di/dt, dv/dt in switch with 50 kHz switching frequency and EMI with

Power stage	Methodology	Additional components	Achieved objective	Reference
Boost	Additional switch of 30 kHz (IGBT)	Switch, inductor, capacitor couple of diode	High efficiency, reduced system size	Gil [1]
Non-isolated DC-DC boost	Active clamp ZCS cell (MOSFET)	Clamping diode	Circulating loss is minimized, reduced voltage stress in diodes, ZCV in all switches, high efficiency	Wu et al. [2]
Bidirectional DC/DC PWM buck-boost	Single common auxiliary circuit	Coupled inductors, single active switch passive elements	Step voltage conversion ratio	Das et al. [3]
Boost	Simple auxiliary resonant circuit	Auxiliary switch, diode, resonant inductor, resonant capacitor	Reduced switching losses, reduced voltage and current stress of the switching devices, easy control	Sang-Hoon Park et al. [4]
Boost	Current transition PWM	Additional cell in main switch, resonant inductor	Reduced di/dt & dv/dt, EMI is reduced, current stress is reduced efficiency improved	Lee et al. [5]
Boost	Zero voltage switching resonant	DC-link in parallel	Voltage stress in switch is minimized, sub-harmonic problem due to DPWM is eliminated	Lee et al. (2001)
Boost	Resonant soft switching technique, power conditioning system, Interleaving	Not required	Reduced switching losses, high efficiency no additional switches are needed for soft switching	Doo-Yong-Jong (1991)
Buck-boost	Zero-voltage switching, zero current switching interleaving technique	Resonant inductor, resonant capacitor, two parasitic capacitors and an auxiliary switch	Voltage and current stress in switches are reduced, input and output current ripples are reduced, size and cost is reduced	Doo-Yong Jung et al. (2001)
Fly back	Single capacitor snubber	Two turnoff snubbers (with inductor, capacitor, diode)	Smooth turnoff of switch, with reduced system complexity	Chen et al. [6]
Fly back	Dual series resonant active clamp technique	Clamp capacitor, auxiliary switch	Switching losses and reverse recovery losses is reduced, high efficiency is obtained	Tseng et al. [7]

Table 1.
Literature summary of soft switching technique.

ZCT-PWM. An additional cell is connected across the main switch reduced di/dt obtained by resonant inductor by soft turn on of main switch.

Doo-Yong Jung et al. [10] connected a photovoltaic power conditioning system to a conventional 1.2 KW ILBC for PV power generation application. The author worked as an improvement in ILBC mentioned in literature in Caris et al. [11] by connecting PVPCS (PV power conditioning system). Interleaved resonant soft switching method is implemented to reduce switching losses. MPPT using PV simulator and current balancing control is done. The work achieved 97.28% efficiency.

Chen et al. [6] three stage interleaving topologies in boost converter handles both voltage and current stress across the switches also get rid of the input andvoutput current ripples in the converter. The 600 W converter works in a couple of operational mode depending on the duty cycle with a switching frequency of 50 kHz and 250 V DC input and 400 W DC output voltage.

Tseng et al. [7] proposed an isolated interleaved buck converter (ILBC) with lossless turn off switches of 50 kHz using a snubber circuit. The snubber is single capacitive turn off snubber connected to 150 V input voltage and 100 V DC output voltage respectively. Work results in less number of components and complexity.

Lee et al. [12] proposed a 400 W fly back converter which is designed in such a way that the series resonance help in PWM, dual series circuit does the ZCS turn off of the diode, active clamping is responsible for magnetizing inductance and ZVS turn on. Thus, high efficiency is obtained by reducing the voltage stress across switch with 100 kHz switching frequency hence reducing switching losses and reverse recovery losses.

Table 1 summarizes the few soft switching topologies discussed in the above Section 1.

Below are the few advantages of this technique that have been witnessed during this literature,

- The semiconductor switches are handled in a smooth fashion during its operation to heavily bring down the switching stress.
- In addition, the switching losses (i.e.) due to di/dt and dv/dt are eradicated to a great extent.
- Obviously, the improved system efficiency brings down the cost of the converter system.
- It is also understood that by soft switching, EMI problems and
- Sub-harmonic problems have also been treated by the authors of the literature.

Therefore, all benefits put together produce a good conversion efficiency to the converter.

It is understood from the above literature that the soft switching is done by introduction of an external or auxiliary circuit in parallel to the main switch to make the turn on and turn off in a smooth fashion without power loss. Therefore, a good conversion efficiency has been obtained for the converter.

2. Review on interleaving technique topologies

Ultimately to extract maximum energy from the solar panel a converter is connected between the panel and the load. If high step converter is connected to the panel then the efficiency will be high. This includes ripple reduction with different conversion stages for different load conditions with different filters. A single-phase boost converter switches due to its extreme duty cycle faced high conduction losses and reverse recovery problem. The voltage stress in the semiconductor switches played a major role in the reduction of efficiency. So, it is evident that efficiency and the cost of the converter are proportional to each other. In addition to that they also suffer high current ripple which does not greatly support MPPT algorithms. Interleaved converters have showed a significance improvement rather than the conventional boost converter which is justified by our previous work. With all the above statements, it has been concluded that the system efficiency is improved due to the fact that interleaving technique reduces the ripple current. The researchers yet have handled many various topologies and few of them are listed below.

Numerous interleaving topologies have been evolved by the researchers of the previous decade, many such topologies have been investigated in this section and their methodologies are summarized in **Table 1** with some cited reference.

Garcia et al. [13] proposed Interleaved double dual boost converter IDDB of 200 KW power and 60 V DC and 360 V DC input and output voltage respectively. It is a non-isolated step up DC-DC converter which gains high voltage and suits electric vehicle, renewable energy, high power application. Six phase IDDB is verified approximately.

Tseng and Huang [14] proposed interleaved converters of 1000 W output power to obtain high step up gain without operating at extreme duty ratio. High step up converter suitable for renewable energy system with a voltage multiplier module. Proposed converter has reduced current stress, reduced input current ripple, and reduced conduction loss, extended life time of the source and improved efficiency. The converter has an Input voltage of 40 V, output of 380 V and efficiency of 97.1% is obtained.

Lai et al. [15] included a forward energy delivering circuit and a voltage doubling circuit with the conventional ILBC to achieve step up ratio in turn high efficiency which is applied to micro grid. Here, interleaving technique is implemented for a high step up converter. Closed loop control of 450 W, 2 paralleled modules with 24 V input and 200 V output is implemented. Module efficiency up to 95.8%. Steady-state analysis has been done.

Kejin et al. (2016) proposed a novel bi-directional three-phase DC/DC converter for high-power application. The proposed converter yields high conversion ratio than other conventional converters. The current and voltage stress of the power switches has been taken care to yield high efficiency.

Akın [16] compared conventional boost converters and interleaved PFC boost converters. They are compared by their power factor (PF), total current harmonic distortion (THD) and total efficiency. ILBC shows an improved efficiency than conventional. Conventional boost converter has 97.8% efficiency and 4.88% THD PF ILBC 98% efficiency and 1.93% THD values.

Table 2 summarizes the few interleaving topologies discussed in the above Section 2.

Thus, from the literature on interleaving techniques we can conclude that interleaving technique is

- The best solution to get rid of input and output current ripple as well as input and output voltage ripple.

- Due to reduced switching loss, conduction loss and voltage stress in the converter switches the system efficiency is greatly improved.

- Since the extreme duty cycle of the semiconductor switches have been minimized, the life time of the switches are increased and in turn resulted in reduced maintenance cost of the converter.

Power stage	Methodology	Achieved objective	Reference
Buck	Two active switches in series, coupling capacitor in power path	Low switching losses, improved step-down conversion ratio, reduced output current ripple	Lee et al. [5]
Non-isolated boost	Interleaved double dual boost (IDDB)	High voltage gain, suitable for high power applications	Garcia et al. [13]
Cuk and boost	Comparison	Cuk-design complexity, boost-high power application, improved switching frequency reduced ripple	Newton (2010)
Boost	High step up gain without operating at extreme duty ratio with voltage multiplier module	Reduced current stress, reduced input current ripple, reduced conduction loss, extended life time of the source and improved efficiency of 97.1% achieved	Tseng and Huang [14]
Boost, interleaved power factor correction boost (PFC ILBC)	Comparison	PFC ILBC-improved power factor (PF), reduced total current harmonic distortion (THD) of nearly 3%, improved efficiency of 0.2% than conventional converter	Akın [16]
Boost	Interleaving, adding forward energy delivering circuit and a voltage doubling circuit to ILBC	High step up ratio in turn high efficiency of 95.8% and also applied to Micro grid	Lai et al. [15]
Boost	Adding coupled inductor to ILBC	Reduces the current stress, current ripple and conduction losses	Tseng and Huang [14]
Boost	Bi-directional three-phase	High conversion ratio, current and voltage stress in switches are reduced	Jin and Liu [17]

Table 2.
Literature summary of interleaving technique.

- Interleaved converters are superior due to its simple structure and low complexity.
- From the above literature, it is also evident that they can serve for wide range of applications.

Though many techniques have been evolved and implemented in the boost converter still lag exists in the research in extracting 100% energy from the solar cell. Interleaving technique being one of the best solutions to get ripple free boosted voltage which results in improved efficiency. But, above all care should be taken in load matching of the converters.

3. Review on special converters

In literature, researchers have compared SEPIC converter, ZETA, Cuk converters with ILBC. SEPIC converter means single ended primary inductance converter and ZETA converter.

Newton [18] applied interleaving concept to the Cuk converter and analysis of system advantages over ILBC is compared. Cuk converter provides isolation without addition of extra active devices. Isolation gives different input and output facility.

Rech et al. (2001) compared different hybrid converters and new hybrid converter is proposed for high power application. The number of level of power conversion is reduced and THD is also reduced.

Yan Li et al. (2014) suggested buck-boost mode is preferred in applications where the input current needs to continuous and the output with less ripples. This gives a long-extended battery life. Non-isolated push pull converters (NIPPC's) are compared. Advantages of NIPPC's are less current stress, high efficiency and good EMC performance. Different types of configuration of push pull converters with single and coupled inductors are compared.

Kosai et al. [19] proposed a silicon carbide (SiC) Power semiconductor technology results in SiC rectifiers and controlled semiconductor devices (JFET and MOSFET), where the power converters can be operated in 200°C.

Qahouq et al. [20] proposed that by using only a single output voltage value, the controller can track the MPP of every solar panel with improved tracking speed, cost effective and also reduced system size. This technique works efficiently on non-uniform loads and mismatching conditions due to partial shading with single voltage sensor.

Darwish et al. [21] discusses about the inherent non-linearity of Cuk converter which is the reason for distortion of output voltage and current. A three phase Cuk converter is proposed. The buck-boost nature of the converter provides flexibility for standalone and grid connected applications.

Hong-Tzer et al. [22] introduced a single phase Cuk bridgeless AC-DC power factor correction rectifier which produces positive output voltage. This converter converts negative output voltage to positive value unlike other traditional topologies. The efficiency is improved by reducing the conduction loss of the power switches and also the power factor is maintained at 0.99 all input and output conditions.

Yifannyu et al. (2016) optimized the power imbalance problem in PV system by zero injection sequence injection method. They also introduced and applied the above-mentioned method in a multilevel-cascaded h-bridge converter to test the feasibility and effectiveness of the introduced method.

Forest et al. [23] is implemented isolating intercell transformer (ICT) in an interleaved converter. Two ways are implemented with coupled inductor or transformer. One is bidirectional converter buck or boost, another is bidirectional converter buck-boost. Advantages of this concept are removal of discrete inductors due which magnetic loss is less, high step up ratio and minimization of switching losses.

Khan et al. [24] proposed a synchronous buck converter with high conversion ratio and reduced system size. Gallium nitride (GaN) field-effect transistors are used in the converter. MPPT is implemented for all irradiation conditions where, maximum power extraction is achieved.

Kaouane et al. [25] obtained maximum energy transfer from generator to load side then other conventional converters by designing a SEPIC converter. High efficiency is obtained also during discontinuous conduction mode using PWM controlled switches.

Table 3 summarizes the few special converters involved in PV Power Generation discussed in the above Section 3.

ILBC yields better efficiency with simple design structure which serves for wide range of applications. Moreover, it is evident that it suffered with less ripples when compared the rest of the two converters with increased switching frequency. And also, they are compact in size though additional components are required for the control of switches. Finally, this section dealt with some other type of special

Converter	Methodology	Summary	Results	Reference
Cuk converter	Comparison of Cuk and ILBC	Isolation gives different input and output facility	Cuk converter provides isolation without addition of extra active devices	Singh et al. (2005)
Hybrid converters	Comparison different hybrid converters and new hybrid converter	Applied in high power application	Number of level is reduced and THD is reduced	Rech et al. [26]
Non-isolated push pull converters (NIPPC's)	Push pull converters with single and coupled inductors are compared	Buck-boost mode is preferred in applications where the input current needs to continuous and the output with less ripples	Long extended battery life, less current stress, high efficiency and good EMC performance	Yan Li et al. (2014)
MPPT controller	Single output voltage value, the controller can track the MPP	Woks efficiently in non-uniform loads and mismatching conditions due to partial shading with single voltage sensor	Improved tracking speed, cost effective and Reduced system size	Jaber Abu Qahouq et al. [20]
Cuk converter	Buck-boost nature	To overcome inherent non-linearity in turn distortion of output voltage and current	Provides flexibility for standalone and grid connected applications	Yang et al. [22]
Single phase Cuk	bridgeless AC-DC rectifier	Converts negative output voltage to positive value unlike	Improved efficiency reduced conduction losses, power factor is maintained at 0.99 all input and output conditions	Yu et al. [27]
Multilevel-cascaded h-bridge converter	Zero injection sequence injection method	To optimize the power imbalance problem in PV system	Reduced switching loss, high conversion efficiency, simple structure	Khan et al. [24]
Synchronous buck converter	Gallium nitride (GaN) field-effect transistors are used in the converter	For all irradiation conditions where, maximum power extraction is achieved	High conversion ratio and reduced system size	Kaouane et al. [25]
SEPIC converter	DCM mode using PWM controlled switches	—	Maximum energy transferred from generator to load side, High efficiency	Simonetti et al
Interleaved converter	Isolating intercell transformer (ICT)	Two ways are implemented with coupled inductor or transformer (buck or boost, bidirectional buck-boost)	Removal of discrete inductors due which magnetic loss is less, high step up ratio and minimization of switching losses	Forest François et al. [23]

Table 3.
Literature summary special converters involved in PV power generation.

converters like Cuk, SEPIC, etc. Here some work on comparison of ILBC's with SEPIC and ZETA converter, pros and cons between them, some new techniques implemented in special converters are also discussed. **Table 3** summarized the aforementioned features.

4. Review on MPPT technique

As we know PV panel is a non-linear device and in order to be able to extract maximum output power despite changing insolation due to partial shading and changing temperature due to weather conditions, maximum power point tracking (MPPT) is required. However, MPPT works based on maximum power transfer theorem which goes like this, whenever source impedance is equal to the load impedance maximum power is delivered to the load. This can be done by adjusting the duty cycle of a DC-DC converter between the source and the load. The researchers have approached various control techniques to achieve this that is listed in **Table 4**.

Doo-Yong-Jong et al. (2012) imposed MPPT to half bridge PV DC-DC boost converter and full bridge inverter applied to grid which adopts variable step-size algorithm. Grid current control is done with 4th order linear IIR filter controller.

Schuck and Pilawa-Podgurski [28] attended the problem faced by PV input interleaved converter with MPPT to work in asymmetric fashion due the difference in load and source combinations. The author proposes a control technique based on and harmonic elimination and ripple reduction under asymmetric source-load conditions.

Femia et al. [29] proposed distributed MPPT rather than the conventional MPPT to overcome the problem of model mismatch and partial shading in PV modules. Though the author showed improvement in the efficiency is a research gap of complexity in the design.

Qin et al. [30] formed an algorithm to extract maximum power from PV modules with differential power processing converters. Fast tracking is achieved with less number of perturbations for long strings.

Park et al. [4] designed a MPPT controller to track maximum power from PV module during fast transient changes due to environmental conditions with good tracking speed and accuracy. The author analyzed different dynamic PV panel data with high number of samples than before which includes different whether conditions.

Pilawa-Podgurski and Perreault [31] proposed an MPPT power converter with digital control technique for Synchronous buck converter. Obtains local and global MPPT in parallel connected PV modules. Direct integration is done hence economical.

Qin and Pilawa-Podgurski [32] proposed a differential power processing architecture. MPPT with minimum communication and no local sensing is ensured. Partial shading condition is considered. The system ensures extraction of more energy which is employed low voltage buck-boost converter. Control method achieves global and local MPPT operation.

Alexander et al. [33] analyzed different MPPT algorithms. Presence of noise in MPPT algorithm leads to deviation MPP steady state. Parameters like slowing tracking and noise are considered which affect the overall efficiency of the system. Optimizations of these system parameters are determined.

Sundareswaran et al. [34] formed artificial bee colony (ABC) algorithm for MPP global maximum power point (GMPP) which enables maximum utilization of solar

System	Problem	Methodology	Objective achieved	Reference
Boost, half-bridge connected to grid	Unbalance in grid current	Repetitive current control method-current controller (repetitive)-based on fourth order linear phase IIR filter. Variable step-size technique	Minimal devices are used, regulates the grid current	Doo-Yong-Jong et al. (2012)
Asymmetric Interleaved Multiphase	Asymmetric conditions due to difference in source and load conditions	Harmonic elimination control technique	Reduced Ripples, works in a symmetric condition, first harmonic ripple component is reduced by 14.8 dB	Schuck and Pilawa-Podgurski [28]
Interleaved	Effect of model mismatch and partial shading in PV module	Distributed MPPT	Improved efficiency, reliable	Femia et al. [29]
Series connected	To maximize the power extraction from series connected PV module	Differential power processing algorithm	Fast tracking without extra hardware, reduced no. of perturbations in each step, suitable for long sub module strings	Qin et al. [30]
DC-DC, DC-AC	Rapid transient changes in PV output power	MPPT controller design considering field measurements of transient effects	Improved tracking speed and accuracy—captures dynamic PV panel data of high temporal resolution	Park et al. [4]
Synchronous buck converter	To increase of the energy captured during partial shading conditions. To reduce the overall cost	Sub module integrated digital control technique	Ensures both local and global MPPT, direct integration is done hence economical	Pilawa-Podgurski and Perreault [31]
Buck-boost converter	extraction of more energy from PV applications	Differential power processing architecture (MPPT with minimum communication and no local sensing)	Global and local MPPT is obtained	Qin and Pilawa-Podgurski [32]
PV power generation	MPPT under partial shading conditions	ABC algorithm	Ensures fast tracking characteristics, improved energy saving capability, improved revenue generation	Alexander et al. [33]

System	Problem	Methodology	Objective achieved	Reference
Bidirectional DC-DC converter	MPPT for standalone power generation	Parallel MPPT technique	Reduced loss in power converter, improved efficiency, reduced number of power processing stages	Sundareswaran et al. [34]
Standalone hybrid system (solar and wind)	To achieve fast, better transient response and pitch angle of wind turbine	Neural-Network-Based MPPT control, radial basis function network (RBFN), Elman Neural Network (ENN)	Fast and improved transient response	Subudhi and Pradhan [35]
PV system	To minimize the settling time.	FPGA platform to obtain real time weather conditions, optimization of perturbation period in P&O algorithm, the settling time is calculated by dual kalman filter	Settling time is minimized	Roger Gules et al. (2008)
Buck converter	To improve solar energy conversion efficiency	Fuzzy-PID controller	Achieving high MPP under various environmental conditions	Whei-Min Lin Member et al. (2011)
PV system	To obtain maximum power transfer to load	Two phase tracking INC MPPT	Maximum power transfer	Jiang et al. [36]
PV system	To attain GMPP	GMPPT technique with a simulated annealing (SA)	Less design complexity, less number of parameters per iteration in P&O algorithm	Sarah and Haque [37]
Grid connected PV system	To identify maximum power point (MPP) rapidly	Modern predictive control (MPC)	Dynamic response of the system is improved	Shadmand et al. [38]
PV system	Uncertainty due to rapidly changing weather condition and variable load condition	Double integral sliding mode controller (DISMC) MPPT system	Easy to control where, switching frequency is kept constant	Raseswari et al. [39]
PV system	To avoid steady-state oscillation, to improve tracking efficiency	Gray wolf optimization method for MPPT tracking	Better transient response	Seyed Mahmoudian et al.

Table 4.
MPPT techniques imposed in solar powered converters.

energy under varying atmospheric conditions. Two different PV configurations under different shading patterns are tested. This ensures fast tracking characteristics, improves energy saving capability, increased revenue generation when compared with other energy MPPT techniques.

Subudhi and Pradhan [35] reviewed MPPT techniques developed until January 2012. Analysis based number of control variables, types of control strategy, types of circuitry for PV systems and application based on practical and commercial. Nearly 26 variety of MPPT technique has been compared.

Gules et al. [40] proposed a bidirectional DC-DC Converter with parallel MPPT for standalone power generation. The power converter works as multifunctional MPPT circuit, battery charger, battery regulator and as step up converter. Parallel MPPT helps in reducing losses in the power converter in turns improves efficiency also reduces the number of power processing stages. Same series connected MPPT algorithms can be used.

Lin et al. [41] constructed a standalone hybrid system consists of solar power, wind power, diesel engine and intelligent power controller. Radial basis function network (RBFN) is used to achieve fast and transient response and pitch angle of wind turbine and Elman Neural Network (ENN) is used to achieve MPPT.

Jiang et al. [36] proposed a grid connected boost half bridge PV micro inverter system which uses only minimal device. Dynamic response is improved when load and solar irradiance were changed rapidly. Variable step size MPPT is adopted.

Ricco et al. [42] used FPGA platform to MPPT algorithms acquire its parameters to the real-time weather conditions. The settling time is minimized. The proposed controller presents optimizing the perturbation period of the P&O algorithm the settling time can be calculated by a dual kalman filter.

Dounis et al. [43] proposed that Maximum power point tracking (MPPT) with Fuzzy-PID controller to regulate to output power of PV system for a DC/DC buck converter. The proposed system improves the solar energy conversion efficiency. Fuzzy-PID is highly effective in achieving high MPP under various environmental conditions.

Hsieh et al. [44] proposed two phase tracking INC MPPT which either constant-frequency with variable-duty cycle (CFVD) or variable-frequency with constant duty control(VFCD) to perform with reference to a commonly known threshold-tracking zone (TTZ) thus obtained maximum the power transfer to load at MPP.

Wang et al. [45] identified that the mismatches the PV system affects the major portion of energy production. Then, proposed a MPPT system which is simple with unified output.

Lyden et al. (2015) proposed that the global maximum power point tracking (GMPPT) technique with a simulated annealing (SA) for PV energy system which experiences the partial shading condition (PSC). This attains GMPP or near GMPP. Less design complexity when compared to P&O technique and less number of parameters per iteration to iteration. Shadmand et al. [38] proposed a MPPT technique called as modern predictive control (MPC) for a grid connected PV system. The technique identifies the Maximum Power Point (MPP) rapidly than other earlier techniques. The dynamic response of the system is improved than other conventional techniques.

Pradhan and Subudhi [39] handled the uncertainty problem due to rapidly changing weather condition and variable load condition in a PV system by implementing double integral sliding model controller (DISMC) MPPT system. The system is easy to control where, switching frequency is kept constant. Mohanty et al. [46] used Gray wolf optimization method for MPPT tracking. The limitations such as steady-state oscillation, lower tracking efficiency and transient condition has been fixed in this method. Where, our other traditional methods such as P&O, improved P&O failed to do so.

Table 4 summarizes the few MPPT techniques imposed in solar powered converters discussed in the above Section 4.

It is clearly evident from the above literature review that various problems have been attended and solved by implementing MPPT controller to a system. To summarize some of the key problems addressed,

- Uncertainty due to partial shading and changing weather conditions.
- Asymmetric source-load mismatch, model mismatch especially in interleaved converters.
- Rapid transient changes in PV power generation system.
- Steady state oscillations are avoided by minimizing the settling time.
- Unbalanced grid current is handled.

The MPPT techniques used by the researchers in the previous decades have been discussed in the above section. The Problem, objectives, various techniques and advantages of MPPT imposed PV system are summarized and discussed in **Table 4**. The MPPT applied PV power generation system have witnessed the following advantages, which is understood from the literature.

- High conversion efficiency because of maximum power extraction especially in series connected models.
- Reduced ripples and harmonic distortions in the system.
- Fast maximum power tracking in turn better tracking efficiency.

Supports wide range of applications, which includes stand-alone power generation and also in pitch angle control for wind turbines.

5. Review on controller for ILBC

Since the boost converters are nonlinear in nature, the controller maintains the stability during the source-load variation. In order to balance the fast-transient response, it is vital to select an appropriate controller to overcome over shoot damping and achieve the steady state response. Different types of controllers used by the researchers so far till 2016 have been discussed here. The extended summary is displayed in **Table 5**.

Zdravko et al. [51] proposed a sensor less multiphase average current controller which a self-tuning current estimator. The average inductor current is calculated with the duty cycle, input and output currents, based on these parameters the active phase gets automatically changed.

Cheng et al. [47] introduced Hybrid ripple concept for ripple-based control which may suffer from large jitter issues and instability. It is achieved by feed backing both inductor current ripple and compensated external ramp to the modulator. Reduce jitter and improved stability is achieved. Fast load transient performance with the proposed control scheme of controller is achieved. Research gap: system becomes more complicated with two different ramp compensations. Wang and Li [48] developed a RPC (ratio pre-setting controller) method to enhance current

Controller	Technique	Methodology	Summary	Results	Reference
Multiphase average current controller	Self-tuning current estimator	Average inductor current is calculated with the duty cycle, input and output currents	Using these parameters active phase gets automatically changed	Sensor less controller	ZdravkoLukić (2009)
Ripple-based control	Hybrid ripple concept	Feed backing both Inductor current ripple and compensated external ramp to the modulator	Research gap-system becomes more complicated with two different ramp compensations	Reduce jitter and improved stability.	Cheng [47]
Ratio pre-setting controller (RPC)	To enhance current sharing capability	The relationship between phase shift angle and the transformer current is arrived	Three types of three-phase bidirectional DC-DC converters are introduced and other two types are compared	Y-connection type converter evidence less current mismatch due to its current sharing capability	Wang and Li [48]
Power control system	Artificial Neural Network (ANN)	Output voltage is regulated which is received from the PV array from changing ambient temperature and irradiance	Battery as load	Fast settling time and low over shooting	Jiteurtragool et al. [49]
IC controller	IC is embedded with power MOS	Controls the converter as well as takes care of MPPT algorithm	4 Phase ILBC for PV distributed power conversion	Use less number of external components and less overall cost	Pulvirenti et al. [50]
State trajectory controller	State trajectory control	Based on boundary control	Method involves four switching states, and three state equations	Good dynamic response during starting and load change	Rafael Pena-Alzola (2015)

Table 5.
Literature summary of analysis of controllers and implemented applications for PV generation system.

sharing capability. Here, three types of three-phase bidirectional DC-DC converters are introduced and other two types are compared. Where, Y-connection type converter evidence less current mismatch due to its current sharing capability. The relationship between phase shift angle and the transformer current is arrived.

Jiteurtragool et al. [49] proposed a power control system for DC-DC converter using ANN. Output voltage is regulated which is received from the PV array from changing ambient temperature and irradiance. With battery as load hardware implemented with fast settling time and low over shooting. Pulvirenti et al. [50] proposes IC controller 4 Phase ILBC for PV distributed power conversion. The system uses less number of external components and less overall cost is assured. An IC is embedded with power MOS which controls the converter as well as takes care of MPPT algorithm. Sean and Williamson [52] compared various state space model for power amplifiers and concluded there is a considerable difference in the measured frequency response. Three different models are compared in this chapter. Rafael Pena-Alzola et al. (2015) handled instability problems in ILBC's during large bandwidth requirements by state trajectory control which is based on boundary control where, conventional controller failed to do so. The system provides a good dynamic response during starting and load change. The method involves four switching states, and three state equations.

Table 5 summarizes the few analysis of controllers and implemented applications for PV Generation system discussed in Section 5.

Implementation of controller is also being a best method to eliminate ripple. Not only for ripple reduction, the controller also helped to improve the systems overall performance by maintaining good dynamic response.

6. Ripple minimization technique

When input current is equal to the inductor current then the converter works in continuous conduction mode (CCM). But the drawback in this style of working is the diode reverse recovery current. When the inductor current does not match with input current then is said to be working in pulsating mode. In this mode, the input and inductor currents are equal only when it is switched ON. Boost converter works in CCM mode but buck converter works in pulsating mode. So, in pulsating mode of operation the inductor current turns zero when the switch is OFF. Coupled inductor can be a solution for this issue but then integrated magnetic components have caused EMI problems so coupling coefficient should be designed well in advance. So, pulsating type interleaving technique can be adopted to achieve current ripple reduction.

Kim [53] proposed reconfigurable solar converter (RSC) which helps in single stage power conversion. The proposed system is nothing but a single stage, 3 phase grid tie solar PV converter for DC-DC or DC-AC generation used for PV battery application. Numbers of conversion stages are minimized. Therefore, improved efficiency, reduced cost, weight and volume are ensured.

Garcia et al. [54] handled the tolerance in the inductance on the output current and the voltage ripple. The author concluded that Multiphase is advantageous for its filter less input and output. Where, in the other side tolerance in semiconductor switches is undesirable.

Caris et al. [11] proposed a method to calculate the phase-shifts to avoid harmonics generated due to the cell inductors in the interleaved converters. Interleaved topologies used to improve power converters to obtain higher power ratings, higher bandwidth and lower current ripple. On other hand, it suffers with cell inductor tolerance.

Yu Gu et al. (2013) proposed ILBC with RPN (ripple cancelation network). RPN is connected with a normal ILB, it consists of additional inductors, a couple of capacitors and coupled inductor. Input ripple current cancelation with minimum loss and 96% efficiency is achieved.

Caracas et al. [55] implemented solar water pump where, a 3-phase induction motor runs directly from the PV cell without the help of battery. This is a system based on resonant two inductor boost converters (current resonant converter) and full bridge 3 phase voltage source inverter. Also, non-isolated recovery snubber is added with the system in order to control hysteresis loss and duty cycle improvement in order to improve the efficiency. Advantage of resonant two inductor boost converters is high voltage gain and low input current ripple. Prolonged lifetime due to the absence of battery, use of induction motor is more efficient, maintenance free when compared with DC motor.

Peng [56] ILBC with self-balancing capability is designed. Smoother and less distorted power conversion is achieved. More than two level will be convincing to balance DC voltage level without additional circuits. Advantages of system are magnetic less, compact, high efficient, zero EMI, low cost power conversion.

Ramos-Paja et al. [57] analyzed the pre-filter optimal operatinvg conditions where the input ripple is mitigated for better conversion ratio of ILBC. The inductor losses are predicted in terms of efficiency and input current ripple then a pre-filter is designed accordingly.

Problem	Methodology	Results	Remarks	Reference
Load side ripple in parallel multiple in input source	Minimized by increasing the switching frequency	Research gap-increase in switching loss	Can be applied to smart grids	Kim et al. [53]
More number of conversion stage	Reconfigurable solar converter (RSC)	Single stage conversion	Improved efficiency, reduced cost, weight and volume are ensured	Garcia et al. [54]
Tolerance in the inductance on the output current and the voltage ripple	Digital control, multiphase	Filter less input and output	Research gap-undesirable side tolerance in semiconductor switches	Caris et al. [11]
Harmonics in multilevel converter	Calculated phase shift	Higher bandwidth, low current ripple		Gu and Zhang [60]
Input ripple current	Ripple cancelation network (RPN)	Minimum loss and 96% efficiency	Consists of additional inductors, a couple of capacitors and coupled inductor	Peng [56]
Power mismatch	ILBC with self-balancing capability	More smoother and less distorted power conversion	Magnetic less, compact, high efficient, zero EMI, low cost power conversion	Ramos-Paja et al. [57]
Input current ripple mitigation	Pre-filter optimal operating conditions, predicted inductor losses	Better conversion ratio in ILBC		Jiang et al. [59]

Table 6.
Literature summary of ripple minimization techniques in PV power generation.

Tseng et al. [58] included voltage multiplier for managing asymmetrical interleaving condition without changing the duty cycle extremely is proposed. Voltage multiplier is nothing but conventional converter with a coupled inductor. Reduces the current stress, current ripple and conduction losses.

Jiang and Wang [59] discussed about the current ripples in converters and proposed of three phase PWM converter. The author concluded bigger current ripples are generated by discontinuous PWM when compared to space vector PWM. This is mostly suitable for variable switching-frequency control for three phase PWM converters applications.

Table 6 summarizes the few ripple minimization techniques in PV power generation discussed in the above Section 6.

There is high frequency ripple caused by the switching frequency and the low frequency ripple caused by AC load. For the symmetrical 3-phase load, low frequency ripple is not experienced. Rather, the single-phase or unbalanced 3-phase, will experience a double line frequency ripple.

Mainly in interleaving converters source and load mismatch in various conversion stages have occurred. This is handled by the researchers in two ways either,

- By tunning the switching style or,
- By some unique ripple mitigation methods and fixed it with some additional integrated circuit.

In the above two methods the later suffers with switching stress and the former with increased volume of the system size. This method can be considered as bridge to fill the research gap when interleaving technique cannot be applied and considered as limitation. This section discusses about the ripple minimization techniques used so far and **Table 6** exhibits the summary of the same.

7. Power conditioning PV simulator and performance evaluation

Below are some set of references which carried few sophisticated tools for betterment of the PV system that involves

- Evaluation
- monitoring
- As well as rectification

Koran et al. [61] proposed high efficiency and fast transient response time that are essential for PV simulator. PV source simulator-evaluate the power conditioning system and MPPT algorithms, high efficiency photovoltaic source simulator both analog and digital based is proposed.

Kim et al. [62] compared several parameters like efficiency and power factor performance. Several topologies are compared, conventional average current mode control boost PFC, an interleaved boost PFC, a back-to-back bridgeless boost PFC, semi-bridgeless boost PFC, are assessed through loss analysis. Improved power factor correction (PFC) topologies suitable for a high density are identified.

Singh et al. [63] improved power quality by a control strategy for achieving maximum benefits for grid interfacing inverters. Perform a multi-function device by incorporating active power filter functionality. Power converter to inject power generated from RES to the grid.

Rosario et al. (2005) analyzed series—parallel L, Cp, Cs resonant inverter for effect of tolerance, to ensure circuit performance-with no feedback-good repeatability. The author concluded that the capacitor value affect the sensitivity values and done statistical study by Monte-Carlo method.

Roh et al. [64] proposed two phase interleaved critical conduction mode (CRM) power factor correction for boost converter. Variation tolerant phase shifter (VTPS) which ensures accurate 180-degree phase during interleaving operation. This VTPC circuit can be applied to any type of ILBC's.

Stanth et al. (2013) proposed a converter which is connected in parallel with the PV string to improve the energy captured during the shading or power mismatch. The parallel configuration is proposed to handle mismatch of power, its turns off if no mismatch. The converter handles only, mismatch in the adjacent string, turns off during no mismatch, low average power handling and low operational duty cycle these can be considered as research gap.

Oscar et al. [65] developed a multiphase converter with buck converter as base. Natural ZVS is achieved in buck converter the instantaneous inductor current itself will manage the current imbalance in the phases. Better current balance and ZVS in both transitions are achieved but with high current ripple. Higher conduction losses, high number of phase and high current ripple are found as research gap.

Liserre et al. [66] introduced generalized hybrid modulation (GHM) technique to overcome the drawback in the conventional method that these techniques succeed in canceling the sideband harmonics in multilevel voltage only when the DC-link voltages are equal. PSC-PWM technique proposed, a method to obtain the proper shifting angles to avoid the reduction in overall output voltage.

Shibin Qin Andrew et al. (2014) proposed a differential power processing (DPP) which combines micro inverter and DPP converter. The architecture configures DC-DC parallel with PV string. Power yield is improved by overcoming the mismatch in sub module. Since the converters are connected in parallel low power rating is sufficient. Low power losses are evident and suitable for sub module MPPT.

Tolbert et al. [67] introduced a procedure to find all sets of switching times (angles) for each converter in a multilevel converter to obtain desired output voltage. Here, 5th and 7th are eliminated by converting the transcendental equations that identifies the elimination of the harmonics into a set of polynomial equations.

Zhang and Yu [68] investigated on the feature of interleaved PWM DC-DC converter and analyzed their methodology. Practical design is presented such as ripple quantification, zero ripple necessary condition and sufficient condition, fault detection, and waveform shape prediction.

Sangswang and Nwankpa [69] quantified random noise in PWM switching DC-DC boost converter. The presence of noise creates uncertainty in switching time. Performance analysis for the parameters such as stochastic performance index, MPPT and critical energy are done.

Zhang et al. [70] connected pair of additional diode and capacitor/inductor are connected in X-shape to a conventional converter (diode assisted converter). Comparison based on silicon devices, passive components, EMI and efficiency are done. Input current ripple is reduced, improved power conversion and buck-boost capability are ensured.

Ray et al. [71] done a multimode analysis of the ILBC for switching duty cycle from 0 to 1, for both CCM and two DCM modes. This analysis helps designer to optimize the inductor value and coupling factor for different performance values. Also, could optimize the number of power stages for varying input and loads. DCM shows less switching losses.

Li and He [72] reviewed on challenges based on high step up, low cost, high efficiency converters are summarized. To extend the voltage gain, to avoid extreme duty cycle in order to reduce current ripple, to avoid power devices losses by soft switching and to avoid reverse recovery problem of diode are the problems summarized. Limitations of conventional boost converters are analyzed. Parallel connected and non-isolated high step up converters for its low cost and improved efficiency.

Lefevre and Mollov [73] discussed about the asymmetric voltage operation while handling multiphase boost converter. Conversion efficiency of 99% is achieved by introducing a flying-capacitor in the existing topology.

Nigsch et al. [74] proposed a bridgeless single stage PFC converter for efficient solution to fulfill the harmonic standards. Efficiency is improved by removing the input rectifier and providing galvanic isolation by allowing single switch active at a time. Around 95% of efficiency is obtained and single stage approach makes it a low cost high-power density PFC.

Xiong Li et al. (2016) proposed a space vector pulse-width modulation (SVPWM) to regenerate phase current using neutral-point current sensing, scheme. Complementary active vectors replace the zero and offset vectors. Hybrid switching pattern are used, to synthesize the reference voltage vector. Phase currents can be reconstructed using SVPWM scheme.

Yuan-Chuh et al. (2016) achieved desired generator performance by one-cycle voltage regulation (OCVR) for permanent-magnet synchronous generator (SPMSG). Design of voltage controller is simplified and computational time is less when compared to other conventional controllers.

Table 7 summarizes the few power conditioning in PV power generation discussed in the above Section 7.

The power quality monitoring has been done to enhance the system performance, the power quality improvement includes the following,

- Power factor correction
- Power mismatch in multiphase converters
- Fine-tunned control strategy for maximum benefits
- Optimized MPPT algorithms

These set of dynamic response tuning of the PV system are overviewed in this section and power quality issues are also discussed here. The power conditioning, power quality improvement and performance evaluation works of the previous decade has been discussed in the above section and the objective, evaluation parameters, summary of discussions are summarized in **Table 7**.

8. Inference

This research work has compared and identified an efficient and suitable converter extract solar energy from the PV panel in an effective way to obtain high conversion ratio among the conventional converters. The overview exhibited and identified the requirements of a best PV power generation system, where the

Objective	Evaluation parameter/ methodology	Summary	Remarks	Reference
PV source simulator	Efficiency and transient response	Both analog and digital based are proposed	Improved efficiency, fast transient response	Koran et al. [61]
Comparison, power factor correction	Loss analysis of converters—current mode control boost, interleaved boost, buck-boost and semi bridge boost	Efficiency and power factor correction		Kim et al. [62]
Power quality improvement	Performs as a multifunctional device by adding an active power filter	Control strategy to achieve maximum benefits from grid interfacing inverter	Inject power from RES to grid	Mukhtiar Singh et al. [63]
Output current sensitivity analysis	Effect of tolerance, performance without feedback	Capacitor values effect the sensitivity value	Good repeatability	Rosario et al. (2005)
Power factor correction	Variation tolerant phase shifting technique (VTPS), critical conduction mode technique (CRM)	Two phase interleaved converters	Accurate 180-degree phase shift and can be applied to any type of ILBC's	Roh et al. [64]
Improvement of dynamic response	Resonant switched capacitor converter	Converter imposed in parallel with PV string	Research gap-handles only adjacent string mismatch, turns off during mismatch, low average power handling and low operational duty cycle	Stanth et al. (2013)
Current imbalance in the phases	Current self-balance mechanism in CCM mode	Multiphase buck converter	Research gap-high conduction losses, more no of phases, high current ripple	Oscar et al. [65]
Sideband harmonics in multilevel voltage, multilevel phase shifting in non-equal DC link	Generalized hybrid modulation technique, carrier PWM technique	To obtain exact phase shifting angle	Reduction in overall output power is avoided	Liserre et al. [66]

Objective	Evaluation parameter/ methodology	Summary	Remarks	Reference
Mismatch in sub-module DC-DC parallel PV string	Differential power processing (DPP)	Combines micro inverter and DPP converter	Less power is only required since connected in parallel	Shibin Qin, Andrew et al. (2014)
Harmonic elimination	To find switching times (angles)	In multilevel converters where, non-equal DC source	5th and 7th harmonics are eliminated	Tolbert et al. [67]
Ripple quantification, zero ripple necessary condition	Analysis of interleaved DC-DC PWM converter	DC-DC converter	Ripple minimization, fault detection, wave shape prediction	Zhang and Yu [68]
Uncertainty switching time due to noise	Analysis of critical energy, stochastic performance index	PWM switching DC-DC boost converter		Anawach Sangswang and Nwankpa [69]
To optimize the number of power stages for varying input and loads	Multimode analysis of the ILBC for switching duty cycle	Multimode analysis of the ILBC	Helps to optimize the inductor value and coupling factor for different performance values	Ray et al. [71]
Asymmetric voltage operation	Flying capacitor is introduced	Multiphase boost converter	99% conversion efficiency	Li and He [72]
To fulfill harmonic standards	Removing the input rectifier and providing galvanic isolation by allowing single switch active at a time	Bridgeless single stage PFC converter	95% of efficiency is obtained, low cost high-power density PFC	Guillaume Lefevre et al. (2010)
To synthesize the reference voltage vector	Space vector pulse-width modulation (SVPWM)	—	—	Nigsch et al. [74]
Performance enhancement	One-cycle voltage regulation (OCVR)	Permanent-magnet synchronous generator (SPMSG)	Simplified design of voltage controller and less computational time	Li et al. [75]

Table 7.
Literature summary of power conditioning in PV power generation.

investigation involves the work till 2016. The above survey vigorously focussed on the analysis and requirements of an efficient PV power generation system. In order to yield the maximum energy from the PV panel which is highly nonlinear in nature, in spite of its heavy capitation cost and low conversion efficiency a well-developed PV power generation system is essential.

The boost converter which is intended between the PV panel and the load plays a major role in this process of extracting maximum energy. It should boost and regulate the voltage from the source, should have less switching losses semiconductor switches, reduced ripples, should act as an isolation between source and the load and finally needs to be free from EMI. These forth-coming conclusions are derived from the above bunch of literatures that can be considered while designing a PV power generation system.

- Si (crystalline) is considered as the best technology which produces nearly 24.7% of efficiency when compared with rest of the evolved solar cell technologies. Where, in this technology the electron-hole pair recombination seems to less than any other technology.

- Though, soft switching technique fetched a considerable improved status in conversion efficiency there are few limitations, therefore care should be taken to reduce the size and cost of the system due to extra external circuit for control of switches.

- Though, interleaving technique is best solution to get rid of input and output current ripple without the extra external circuit which is considered as the drawback of soft switching technique mentioned above, care should be taken for the source load mismatch.

- In order to overcome the above-mentioned limitation of source load mismatching in interleaved converters maximum power point tracking MPPT can be enabled to the system. With MPPT reduced ripples and harmonic distortions in the system high conversion efficiency is obtained. Care should be taken to choose the best MPPT technique.

- There is high frequency ripple caused by the switching frequency and low frequency ripple caused by AC load. For the symmetrical 3-phase load, no low frequency ripple occurs. For single-phase or unbalanced 3-phase, there will be a double line frequency ripple. In this survey, high frequency ripple caused by switching frequency is concentrated. This is handled by the researchers in two ways either by tuning the switching style or by some unique ripple mitigation methods and fixed it with some additional integrated circuit.

- Implementation of controller is also a best method to eliminate ripple. Not only for ripple reduction, the controller also helps to improve the systems overall performance by maintaining good dynamic response.

- ILBC is better than the other special purpose converters like ZETA and SEPIC in terms of its system efficiency, compact size, ripple reduction, design simplicity.

The survey mainly focused on the requirements to build an efficient PV system with extraction of maximum energy and better dynamic response. The general discussion on objectives, techniques and advantages of each sections were also exhibited. The detail requirements of different constraint are presented in a systematic way in order to facilitate the reader to find a better solution for their application and constraints. It is essential to enhance these methodologies used so far and provide sufficient guidelines for selection criteria to support the researchers of PV power generation to opt and design a best PV architecture.

Author details

M.L. Bharathi
Sathyabama Institute of Science and Technology, Chennai, India

*Address all correspondence to: bharathiml15@gmail.com

References

[1] Cha G-R, Park S-H, Won C-Y, Jung Y-C, Song S-H. High efficiency soft switching boost converter for photovoltaic system. In: Power Electronics and Motion Control Conference, 2008. EPE-PEMC 2008. 13th. IEEE; 2008. pp. 383-387

[2] Wu X, Zhang J, Ye X, Qian Z. Analysis and derivations for a family ZVS converter based on a new active clamp ZVS cell. IEEE Transactions on Industrial Electronics. 2008;**55**(2):773-781

[3] Das P, Mousavi A, Moschopoulos G. A novel ZVS-PWM DC-DC converter for bidirectional applications with steep conversion ratio. In: Energy Conversion Congress and Exposition, ECCE 2009. IEEE. 2009. pp. 2030-2036

[4] Park S-H, Cha G-R, Jung Y-C, Won C-Y. Design and application for PV generation system using a soft-switching boost converter with SARC. IEEE Transactions on Industrial Electronics. 2010;**57**(2):515-522

[5] Lee I-O, Cho S-Y, Moon G-W. Interleaved buck converter having low switching losses and improved step-down conversion ratio. IEEE Transactions on Power Electronics. 2012;**27**(8):3664-3675

[6] Chen Y-T, Shiu S-M, Liang R-H. Analysis and design of a zero-voltage-switching and zero-current-switching interleaved boost converter. IEEE Transactions on Power Electronics. 2012;**27**(1):161-173

[7] Tseng SY, Chang G-K, Chen PF, Peng ST, Fan SY. Flyback converter using single-capacitor snubber for auto-tuning fan speed system. International Journal of Power Electronics. 2010;**2**(4):428-441

[8] Amorndechaphon D, Premrudeepreechacharn S, Higuchi K. Design and analysis of high efficiency DC-DC boost converter with active resonant technique for small grid-connected PV systems. In: TENCON 2010-2010 IEEE Region 10 Conference. 2010. pp. 1192-1197

[9] Lee D-Y, Lee M-K, Hyun D-S, Choy I. New zero-current-transition PWM DC/DC converters without current stress. IEEE Transactions on Power Electronics. 2003;**18**(1):95-104

[10] Jung D-Y, Ji Y-H, Park S-H, Jung Y-C, Won C-Y. Interleaved soft-switching boost converter for photovoltaic power-generation system. IEEE Transactions on Power Electronics. 2010;**26**(4):1137-1145

[11] Caris MLA, Huisman H, Schellekens JM, Duarte JL. Generalized harmonic elimination method for interleaved power amplifiers. In: IECON 2012-38th Annual Conference on IEEE Industrial Electronics Society. 2012. pp. 4979-4984

[12] Lee J-J, Kwon J-M, Kim E-H, Kwon B-H. Dual series-resonant active-clamp converter. IEEE Transactions on Industrial Electronics. 2008;**55**(2):699-710

[13] Garcia Fellipe S, Pomilio JA, Spiazzi G. Modeling and control design of the interleaved double dual boost converter. IEEE Transactions on Industrial Electronics. 2013;**60**(8):3283-3290

[14] Tseng K-C, Huang C-C. High step-up high-efficiency interleaved converter with voltage multiplier module for renewable energy system. IEEE Transactions on Industrial Electronics. 2014;**61**(3):1311-1319

[15] Lai C-M, Pan C-T, Cheng M-C. High-efficiency modular high step-up interleaved boost converter for

DC-microgrid applications. IEEE Transactions on Industry Applications. 2012;**48**(1):161-171

[16] Akin B. Comparison of conventional and interleaved PFC boost converters for fast and efficient charge of Li-ion batteries used in electrical cars. In: International Conference on Power and Energy Systems. Vol. 13. 2012. pp. 499-504

[17] Jin K, Liu C. A novel PWM high voltage conversion ratio bidirectional three-phase DC/DC converter with Y–Δ connected transformer. IEEE Transactions on Power Electronics. 2016;**31**(1):81-88

[18] Newton A, Green TC, Andrew D. AC/DC power factor correction using interleaved boost and Cuk converters. 2000. pp. 293-298

[19] Kosai H, Scofield J, McNeal S, Jordan B, Ray B. Design and performance evaluation of a 200°C interleaved boost converter. IEEE Transactions on Power Electronics. 2013;**28**(4):1691-1699

[20] Qahouq JA, Jiang Y, Huang W. Load-voltage-based single-sensor MPPT controller for multi-channel PV systems. In: Applied Power Electronics Conference and Exposition (APEC), 2014 Twenty-Ninth Annual IEEE. 2014. pp. 3451-3454

[21] Darwish A, Holliday D, Ahmed S, Massoud AM, Williams BW. A single-stage three-phase inverter based on Cuk converters for PV applications. IEEE Journal of Emerging and Selected Topics in Power Electronics. 2014;**2**(4):797-807

[22] Yang H-T, Chiang H-W, Chen C-Y. Implementation of bridgeless Cuk power factor corrector with positive output voltage. IEEE Transactions on Industry Applications. 2015;**51**(4):3325-3333

[23] Forest F, Meynard TA, Labouré E, Gelis B, Huselstein J-J, Brandelero JC. An isolated multicellintercell transformer converter for applications with a high step-up ratio. IEEE Transactions on Power Electronics. 2013;**28**(3):1107-1119

[24] Khan O, Xiao W, Zeineldin HH. Gallium-nitride-based submodule integrated converters for high-efficiency distributed maximum power point tracking PV applications. IEEE Transactions on Industrial Electronics. 2016;**63**(2):966-975

[25] Kaouane M, Boukhelifa A, Chériti A. Design of a synchronous sepic DC-DC converter for a stand-alone photovoltaic system. In: Electrical and Computer Engineering (CCECE), IEEE 28th Canadian Conference on. 2015. pp. 870-874

[26] Rech C, Pinheiro H, Grundling HA, Hey HL, Pinheiro JR. Analysis and comparison of hybrid multilevel voltage source inverters. In: Power Electronics Specialists Conference, 2002. Pesc 02. 2002 IEEE 33rd Annual. Vol. 2. 2002. pp. 491-496

[27] Yu Y, Konstantinou G, Hredzak B, Agelidis VG. Power balance optimization of cascaded H-bridge multilevel converters for large-scale photovoltaic integration. IEEE Transactions on Power Electronics. 2016;**31**(2):1108-1120

[28] Schuck M, Pilawa-Podgurski RCN. Ripple minimization through harmonic elimination in asymmetric interleaved multiphase DC–DC converters. IEEE Transactions on Power Electronics. 2015;**30**(12):7202-7214

[29] Femia N, Lisi G, Petrone G, Spagnuolo G, Vitelli M. Distributed maximum power point tracking of photovoltaic arrays: Novel approach and system analysis. IEEE Transactions on Industrial Electronics. 2008;**55**(7):2610-2621

[30] QinS,CadyST,Dominguez-GarciaAD, Pilawa-Podgurski RCN. A distributed approach to maximum power point tracking for photovoltaic submodule differential power processing. IEEE Transactions on Power Electronics. 2015;**30**(4):2024-2040

[31] Pilawa-Podgurski RCN, Perreault DJ. Submodule integrated distributed maximum power point tracking for solar photovoltaic applications. IEEE Transactions on Power Electronics. 2013;**28**(6):2957-2967

[32] Qin S, Pilawa-Podgurski RCN. Sub-module differential power processing for photovoltaic applications. In: Applied Power Electronics Conference and Exposition (APEC), Twenty-Eighth Annual IEEE. 2013. pp. 101-108

[33] Latham Alexander M, Pilawa-Podgurski R, Odame KM, Sullivan CR. Analysis and optimization of maximum power point tracking algorithms in the presence of noise. IEEE Transactions on Power Electronics. 2013;**28**(7):3479-3494

[34] Sundareswaran K, Sankar P, Nayak PSR, Simon SP, Palani S. Enhanced energy output from a PV system under partial shaded conditions through artificial bee colony. IEEE Transactions on Sustainable Energy. 2015;**6**(1):198-209

[35] Subudhi B, Pradhan R. A comparative study on maximum power point tracking techniques for photovoltaic power systems. IEEE Transactions on Sustainable Energy. 2013;**4**(1):89-98

[36] Jiang S, Cao D, Li Y, Peng FZ. Grid-connected boost-half-bridge photovoltaic microinverter system using repetitive current control and maximum power point tracking. IEEE Transactions on Power Electronics. 2012;**27**(11):4711-4722

[37] Lyden S, Haque ME. A simulated annealing global maximum power point tracking approach for PV modules under partial shading conditions. IEEE Transactions on Power Electronics. 2016;**31**(6):4171-4181

[38] Shadmand MB, Mosa M, Balog RS, Rub HA. Maximum power point tracking of grid connected photovoltaic system employing model predictive control. In: Applied Power Electronics Conference and Exposition (APEC), 2015 IEEE. 2015. pp. 3067-3074

[39] Pradhan R, Subudhi B. Double integral sliding mode MPPT control of a photovoltaic system. IEEE Transactions on Control Systems Technology. 2016;**24**(1):285-292

[40] Gules R, Pacheco JDP, Hey HL, Imhoff J. A maximum power point tracking system with parallel connection for PV stand-alone applications. IEEE Transactions on Industrial Electronics. 2008;**55**(7):2674-2683

[41] Lin W-M, Hong C-M, Chen C-H. Neural-network-based MPPT control of a stand-alone hybrid power generation system. IEEE Transactions on Power Electronics. 2011;**26**(12):3571-3581

[42] Ricco M, PatrizioManganiello EM, Petrone G, Spagnuolo G. FPGA-based implementation of dual kalman filter for PV MPPT applications. IEEE Transactions on Industrial Informatics. 2015

[43] Dounis AI, Stavrinidis S, Kofinas P, Tseles D. Fuzzy-PID controller for MPPT of PV system optimized by Big Bang-Big Crunch algorithm. In: Fuzzy Systems (FUZZ-IEEE), IEEE International Conference on. 2015. pp. 1-8

[44] Hsieh G-C, Hsieh HI, Tsai C-Y, Wang C-H. Photovoltaic power-increment-aided incremental-conductance MPPT

with two-phased tracking. IEEE Transactions on Power Electronics. 2013;**28**(6):2895-2911

[45] Wang F, Xinke W, Lee FC, Wang Z, Kong P, Zhuo F. Analysis of unified output MPPT control in subpanel PV converter system. IEEE Transactions on Power Electronics. 2014;**29**(3):1275-1284

[46] Mohanty S, Subudhi B, Ray PK. A new MPPT design using grey wolf optimization technique for photovoltaic system under partial shading conditions. IEEE Transactions on Sustainable Energy. 2016;**7**(1):181-188

[47] Cheng K-Y, Yu F, Yan Y, Lee FC, Mattavelli P, Wu W. Analysis of multi-phase hybrid ripple-based adaptive on-time control for voltage regulator modules. In: Applied Power Electronics Conference and Exposition (APEC), Twenty-Seventh Annual IEEE. 2012. pp. 1088-1095

[48] Wang Z, Li H. Three-phase bidirectional dc-dc converter with enhanced current sharing capability. In: Energy Conversion Congress and Exposition (ECCE). 2010. pp. 1116-1122

[49] Jiteurtragool N, Wannaboon C, San-Um W. A power control system in DC-DC boost converter integrated with photovoltaic arrays using optimized back propagation artificial neural network. In: Knowledge and Smart Technology (KST), 5th International Conference on. 2013. pp. 107-112

[50] Pulvirenti F, La Scala A, Ragonese D, D'Souza K, Tina GM, Pennisi S. 4-phase interleaved boost converter with IC controller for distributed photovoltaic systems. IEEE Transactions on Circuits and Systems I: Regular Papers. 2013;**60**(11):3090-3102

[51] Lukic Z, Ahsanuzzaman SM, Prodic A, Zhao Z. Self-tuning sensor less digital current-mode controller with accurate current sharing for multi-phase dc-dc converters. In: Applied Power Electronics Conference and Exposition, APEC 2009. Twenty-Fourth Annual IEEE. 2009. pp. 264-268

[52] Smithson Sean C, Williamson SS. A unified state-space model of constant-frequency current-mode-controlled power converters in continuous conduction mode. IEEE Transactions on Industrial Electronics. 2015;**62**(7):4514-4524

[53] Kim H, Parkhideh B, Bongers TD, Gao H. Reconfigurable solar converter: A single-stage power conversion PV-battery system. IEEE Transactions on Power Electronics. 2013;**28**(8):3788-3797

[54] Garcia O, Zumel P, De Castro A, Cobos JA. Effect of the tolerances in multi-phase dc-dc converters. In: Power Electronics Specialists Conference, PESC'05. IEEE 36th. 2005. pp. 1452-1457

[55] Caracas JVM, De Carvalho Farias G, Teixeira LFM, De Souza Ribeiro LA. Implementation of a high-efficiency, high-lifetime and low-cost converter for an autonomous photovoltaic water pumping system. IEEE Transactions on Industry Applications. 2014;**50**(1):631-641

[56] Peng Fang Z. A generalized multilevel inverter topology with self-voltage balancing. IEEE Transactions on Industry Applications. 2001;**37**(2):611-618

[57] Ramos-Paja CA, Carrejo C, Saavedra A, Ramirez I, Paniagua L, Giral R, et al. A ripple-mitigating pre-filter based on interleaved DC-DC boost converters. In: IECON 2010-36th Annual Conference on IEEE Industrial Electronics Society. 2010. pp. 2771-2777

[58] Tseng K-C, Huang C-C, Shih W-Y. A high step-up converter with a voltage

multiplier module for a photovoltaic system. IEEE Transactions on Power Electronics. 2013;**28**(6):3047-3057

[59] Jiang D, Wang F. Current-ripple prediction for three-phase PWM converters. IEEE Transactions on Industry Applications. 2014;**50**(1):531-538

[60] Yu G, Zhang D. Interleaved boost converter with ripple cancellation network. IEEE Transactions on Power Electronics. 2013;**28**(8):3860-3869

[61] Koran A, LaBella T, Lai J-S. High efficiency photovoltaic source simulator with fast response time for solar power conditioning systems evaluation. IEEE Transactions on Power Electronics. 2014;**29**(3):1285-1297

[62] Yun-Sung K, Sung W-Y, Lee B-K. Comparative performance analysis of high density and efficiency PFC topologies. IEEE Transactions on Power Electronics. 2014;**29**(6):2666-2679

[63] Singh M, VinodKhadkikar AC, Varma RK. Grid interconnection of renewable energy sources at the distribution level with power-quality improvement features. IEEE Transactions on Power Delivery. 2011;**26**(1):307-315

[64] Roh Y-S, Moon Y-J, Park J, Yoo C. A two-phase interleaved power factor correction boost converter with a variation-tolerant phase shifting technique. IEEE Transactions on Power Electronics. 2014;**29**(2):1032-1040

[65] García O, Zumel P, de Castro A, Alou P, Cobos JÉA. Current self-balance mechanism in multiphase buck converter. IEEE Transactions on Power Electronics. 2009;**24**(6):1600-1606

[66] Liserre M, Monopoli VG, Dell'Aquila A, Pigazo A, Moreno V. Multilevel phase-shifting carrier PWM technique in case of non-equal dc-link voltages. In: IEEE Industrial Electronics, IECon 2006-32nd Annual Conference on. 2006. pp. 1639-1642

[67] Tolbert LM, Chiasson J, McKenzie K, Zhong D. Elimination of harmonics in a multilevel converter with nonequal DC sources. In: Applied Power Electronics Conference and Exposition, APEC'03. Eighteenth Annual IEEE. Vol. 1. 2003. pp. 589-595

[68] Zhang S, Xiaoyan Y. A unified analytical modeling of the interleaved pulse width modulation (PWM) dc–dc converter and its applications. IEEE Transactions on Power Electronics. 2013;**28**(11):5147-5158

[69] Sangswang A, Nwankpa CO. Noise characteristics of DC-DC boost converters: Experimental validation and performance evaluation. IEEE Transactions on Industrial Electronics. 2004;**51**(6):1297-1304

[70] Zhang Y, Liu J, Ma X, Feng J. Comparison of conventional DC-DC converter and a family of diode-assisted DC-DC converter in renewable energy applications. Journal of Power Electronics. 2014;**14**(2):203-216

[71] Ray B, Kosai H, McNeal S, Jordan B, Scofield J. A comprehensive multi-mode performance analysis of interleaved boost converters. In: Energy Conversion Congress and Exposition (ECCE). 2010. pp. 3014-3021

[72] Li W, He X. Review of non-isolated high-step-up DC/DC converters in photovoltaic grid-connected applications. IEEE Transactions on Industrial Electronics. 2011;**58**(4):1239-1250

[73] Lefevre G, Mollov SV. A soft-switched asymmetric flying-capacitor boost converter with synchronous rectification. IEEE

Transactions on Power Electronics. 2016;**31**(3):2200-2212

[74] Nigsch S, Cuk S, Schenk K. Analysis, modeling and design of a true bridgeless single stage PFC with galvanic isolation. In: Applied Power Electronics Conference and Exposition (APEC), 2015 IEEE. 2015. pp. 469-476

[75] Li X, Dusmez S, Akin B, Rajashekara K. A new SVPWM for the phase current reconstruction of three-phase three-level T-type converters. IEEE Transactions on Power Electronics. 2016;**31**(3):2627-2637

Section 3

Fault Detection and Control in Renewable Energy Systems

Chapter 9

A New Control Strategy for Photovoltaic System Connected to the Grid via Three-Time-Scale Singular Perturbation Technique with Performance Analysis

Youssef Mchaouar, Abdelmajid Abouloifa, Ibtissam Lachkar and Mohammed Fettach

Abstract

This chapter addresses the problem of controlling single-phase grid-connected photovoltaic system through a full bridge inverter with L-filter. The control objectives are threefold: (i) forcing the voltage in the output of photovoltaic panel to track a reference. This reference has been obtained from the maximum power point tracking strategy; (ii) guaranteeing a tight regulation of the DC-link voltage; and (iii) ensuring a satisfactory power factor correction (PFC) at the grid such as the currents injected must be sinusoidal with the same frequency and the same phase as the grid voltage. The considered control problem entails several difficulties including: (i) the high dimension and strong nonlinearity of the system; (ii) the changes in atmospheric conditions. The problem is dealt with by designing a synthesized nonlinear multi-loop controller using singular perturbation technique, in which a three-time-scale dynamics is artificially induced in the closed-loop system. A formal analysis based on the three-time-scale singular perturbation technique and the averaging theory is developed to proved that all control objectives are asymptotically achieved up to small harmonic errors (ripples). The performance of the proposed approach and its strong robustness with respect to climate changes are evaluated based on the various simulations results carried out under Matlab/Simulink software.

Keywords: single-phase grid-connected photovoltaic system, nonlinear control, three-time-scale singular perturbation technique, MPPT, power factor correction, averaging theory, stability analysis

1. Introduction

Due to dramatic increase in energy consumption and thrust to reduce carbon and greenhouse gas emissions from the traditional electric power generation systems, photovoltaic (PV) power generators have gained a great popularity in recent years. Indeed, photovoltaic systems produce electric power without harming the environment, transforming a free inexhaustible source of energy, solar radiation, into electricity. Furthermore, the major advantage of the photovoltaic systems is to meet the basic power requirement of non-electrified remote areas, where grid power has not yet reached. Also, there are other advantages such as the declining cost and prices of solar modules. On the other hand, the importance of PV systems in the solar industry makes these systems more efficient and reliable, especially for utility power in distributed generation (DG) at medium and low voltages power systems [1]. All these considerations assure a promising role for PV generation systems in the near future.

On the other hand, many technical problems, such as untimely failures, could be found on electronic systems related in particular to the transfer and conversion of this energy to the network. Today, most conversion systems often suffer from low yields in real production sites. To meet the requirements of the new international standards on expected performance on associated conversion systems, it is important to make a research effort to solve the many control problems associated with the static power converter and bring this area to a degree of sufficient maturity to make them industrial products in their own right. One of the difficulties caused by the use of a photovoltaic conversion chain is focused on the problem of non-perfect control of the chain between the photovoltaic generator itself and the continuous or alternative type of load.

The efficiency of a PV plant is affected mainly by three factors: the efficiency of the PV panel, the efficiency of the static power converter and its control, and the efficiency of the maximum power point tracking (MPPT) algorithm.

PV grid-connected systems represent the most important field applications of solar energy [2–4]. In general, the power converter interface, from PV module (the DC source) to the load or to the grid, consists of two-stage converters: The first-stage DC/DC converter is usually used to boost the PV voltage and to implement the maximum power point technique. While the second stage is used to convert this power into high-quality AC voltage, with power factor correction (PFC) respecting to the power supply grid (i.e. sinusoidal and in phase with the AC supply voltage).

Maximum power point tracking (MPPT) is required to match the PV array power to the environmental changes achieving to extract the maximum power output from a solar cell [5]. To this end, different MPPT techniques have been proposed such as incremental conductance [6], perturbation and observation (P&O) [7, 8], the hill-climbing, and some other special methods, such as neural networks, fuzzy logic technique [9]. Among all available techniques, a simple and effective MPPT of incremental conductance algorithm is applied to attain the maximum power of PV array in different solar irradiance and temperature condition parameters.

In order to provide a stable controller of DC/DC and DC/AC converters, many linear control methods have been proposed using many methods, such as a fuzzy-proportional integral controlled [10], a simple PR controller [11] where the performances have been illustrated by experimental result. However, in both proposed controllers, the problem of maximizing PV power transfer is not accounted for in the controller design. In contrast to linear control, nonlinear approaches can optimize the dynamic performance of system, such as sliding mode [12], fuzzy-sliding mode [13], feedback linearization [14], singular perturbation technique [15], and many others works [16–18]. In light of the previous descriptions, no theoretical analysis is made to formally prove that the closed-loop control performances are actually achieved.

In this chapter, a multi-loop nonlinear controller is designed and developed via singular perturbation technique (Chapter 11 in Refs. [19, 20]), as was shown in Refs. [21, 22], where three-time-scale dynamics is artificially induced in the closed-loop system. The control objectives are threefold: (i) achieving the MPPT for the PV array; (ii) ensuring a tight regulation of the DC-link voltage; and (iii) ensuring a grid connection with unity power factor (PF). These objectives must be met despite changes of the climatic variables (temperature and radiation). A theoretical stability analysis, for the closed-loop system, is provided using the three-time-scale singular perturbation technique [23, 24] and averaging technique (Chapter 10 in Refs. [19, 25]. The three-time-scale analysis allows to construct a suitable composite Lyapunov function candidate for the closed-loop photovoltaic system, and the stability properties of the resulting subsystems are analyzed providing mathematical expressions for the upper bounds of the singularly perturbed parameters.

Compared to previous works, the contribution of the new nonlinear controller enjoys several interesting features including the following:

- Several control objectives are simultaneously taken into account such as: MPPT, DC regulation, and PFC, whereas only some of these objectives have been tackled in previous works [10, 11].
- A theoretical analysis will prove, using three-time-scale singular perturbation and averaging technique, that the desired multiple objectives are achieved. Such a formal analysis was missing in the previous works [12–15].
- The nonlinearity of the controlled system was preserved in the controller design in order to keep all the properties of the studied system, whereas it is partly or totally ignored in previous controllers [14].
- By including of three-time-scale dynamics in the full-order closed-loop system can ensure to achieve desired properties, such as robust zero steady-state error of the reference input realization, desired output performance specifications (overshoot, settling time), and insensitivity of the output transient behavior with respect to parameter variations and external disturbances.

The content of this chapter is outlined as follows: in Section 2, the grid-connected PV system is described and modeled. Section 3 is devoted to the cascade nonlinear controller design and its performances are formally analyzed in Section 4. The global performance of the closed-loop photovoltaic system will be illustrated by numerical simulation using MATLAB/SIMULINK tool in section 5. A conclusion and a reference list end the chapter.

2. System description

This section describes the modeling of photovoltaic system connected to the grid. The power circuit topology used in the proposed single phase grid connected to the photovoltaic array is shown in **Figure 1**. It consists of the following components: (i) a photovoltaic array which consists of an arrangement of N_s-series and N_p-parallel strings; (ii) an input capacitor C_{pv} and a DC-DC boost converter used to increase the voltage level and achieve MPPT for photovoltaic array; (iii) a DC link capacitor C_{dc}; and (iv) a single-phase full-bridge inverter including four power semiconductors with L filter that is used to provide the energy to the grid and ensure power factor correction.

Typical (Ip-Vp) characteristics of solar cells arranged in N_p-parallel and N_s-series can be found in many places (see, e.g. [26]). The PV array module considered in this paper is of type KC200GT. In this chapter, a simple and effective MPPT of

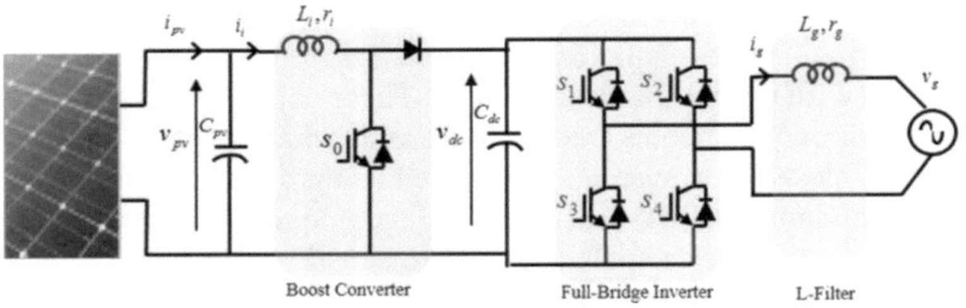

Figure 1.
Single phase grid-connected PV system.

incremental conductance algorithm is applied to attain the maximum power of PV array in different solar irradiance and temperature condition parameters.

By analyzing the circuit and applying the well-known Kirchhoff laws, the system of **Figure 1** can be described by the following set of differential equations:

$$\frac{di_i}{dt} = \frac{v_{pv}}{L_i} - (1 - u_1)\frac{v_{dc}}{L_i} - \frac{r_i}{L_i}i_i \tag{1}$$

$$\frac{dv_{pv}}{dt} = \frac{1}{C_{pv}}\left(i_{pv} - i_i\right) \tag{2}$$

$$\frac{di_g}{dt} = -\frac{v_g}{L_o} - \frac{r_g}{L_g}i_g + u_2\frac{v_{dc}}{L_o} \tag{3}$$

$$\frac{dv_{dc}}{dt} = -\frac{1}{C_{dc}}u_2 i_g + (1 - u_1)\frac{i_i}{C_{dc}} \tag{4}$$

where v_{pv} and i_{pv} are, respectively, the photovoltaic generator voltage and current. v_{dc} and i_{dc} are, respectively, DC link voltage and current. i_i designates the input current chopper, C_{dc} is DC link capacitor, r_i and r_g are, respectively, the equivalent series resistances (ESR) of input inductance L_i and the filter inductance L_g. v_g and i_g are, respectively, the voltage and current of the grid. Here, the grid voltage is defined by $v_g = E_g \sin\left(\omega_g t\right)$, where E_g and ω_g denote the constant amplitude and the constant angular frequency. The switching functions μ_1 and μ_2 are defined by:

$$\mu_1 = \begin{cases} 1 & \text{if} \quad s_0 \quad \text{is} \quad ON \\ 0 & \text{if} \quad s_0 \quad \text{is} \quad OFF \end{cases} \tag{5}$$

$$\mu_2 = \begin{cases} 1 & \text{if} \quad (s_1, s_4) \quad \text{is} \quad ON \quad \text{and} \quad (s_2, s_3) \quad \text{is} \quad OFF \\ -1 & \text{if} \quad (s_1, s_4) \quad \text{is} \quad OFF \quad \text{and} \quad (s_2, s_3) \quad \text{is} \quad ON \end{cases} \tag{6}$$

The instantaneous model (1)–(4) cannot be used directly for the development of continuous control laws since it involves, as input variables, the binary signal μ_1 and μ_2. To overcome this inconvenience, the average model is used [27]. Therefore, the state variables i_i, v_{pv}, i_g, and v_{dc} are replaced by their average values x_1, x_2, x_3, and x_4 over a cutting period. The control inputs u_1 and u_2 denote the average values of μ_1 and μ_2, respectively.

$$\frac{dx_1}{dt} = \frac{1}{L_i}x_2 - \frac{1}{L_i}x_4 - \frac{r_i}{L_i}x_1 + \frac{1}{L_i}x_4u_1 \tag{7}$$

$$\frac{dx_2}{dt} = \frac{i_{pv}}{C_{pv}} - \frac{1}{C_{pv}}x_1 \tag{8}$$

$$\frac{dx_3}{dt} = -\frac{r_g}{L_g}x_3 - \frac{v_g}{L_g} + \frac{1}{L_g}x_4u_2 \tag{9}$$

$$\frac{dx_4}{dt} = -\frac{1}{C_{dc}}x_3\mu_2 - \frac{1}{C_{dc}}x_1u_1 + \frac{1}{C_{dc}}x_1 \tag{10}$$

3. Controller design

3.1 Input inductor current regulation and PFC objectives

3.1.1 Control law design

The first control objective is to enforce the photovoltaic voltage x_2 to track, as closely as possible, the optimal point V'_M (called Regulator 2). However, it is well-known that the boost converter has a non-minimum phase feature. Such an issue is generally dealt with by resorting a cascaded loop design strategy that starts with the input current loop (Regulator 1), as it is shown in **Figure 2**. More specifically, the controller makes the input inductor current x_1 to track a reference signal x_1^*, the latter is determined from (Regulator 2).

In parallel with the input current controller, the network current controller (Regulator 3) will be designed for power factor correction requirement that amounts to forcing the network current x_3 to match the reference signal of the form $x_3^* = \beta \sin(\omega_g t)$. It means that the grid current x_3 should be sinusoidal and in phase with the AC grid voltage v_g, with β is a signal. In fact, the latter is allowed to (and actually will) be time-varying but it must converge to a positive constant value. For both objectives (i.e. the input current x_1 regulation and PFC objectives), let us consider Eqs. (7), (9) in the following form:

$$\dot{X}_{1,3} = M_1U + M_2 \tag{11}$$

where $X_{1,3} = \begin{pmatrix} x_1 \\ x_3 \end{pmatrix}$, $M_1 = \begin{pmatrix} \frac{x_4}{L_i} \\ \frac{x_4}{L_g} \end{pmatrix}$, $M_2 = \begin{pmatrix} -\frac{r_i}{L_i}x_1 + \frac{x_2}{L_i} - \frac{x_4}{L_i} \\ -\frac{r_g}{L_g}x_3 - \frac{v_g}{L_g} \end{pmatrix}$, and $U = \begin{pmatrix} u_1 \\ u_2 \end{pmatrix}$.

Remark 1. Under condition that the DC-link voltage remains all the time positive, it can be showed that $0 < M_{1\min} \le M_1 \le M_{1Max} < \infty$ and $|M_2| \le M_{2Max} < \infty$ are satisfied.

Let us introduce the following current tracking errors:

$$e_i = x_i^* - x_i \quad (i = 1, 3) \tag{12}$$

Then, the reference model can be constructed in the following form

$$\frac{dx_i}{dt} = \frac{e_i}{T_i} \triangleq D'_i(x_i^*, x_i) \quad (i = 1, 3) \tag{13}$$

where T_1 and T_3 denote the time constants and they are selected based on the desired settling time, respectively, for the currents x_1 and x_3. Based on (13), the realization errors of the desired behaviors of $\dot{x}_1$ and $\dot{x}_3$, namely Ω_1 and Ω_3, are given by

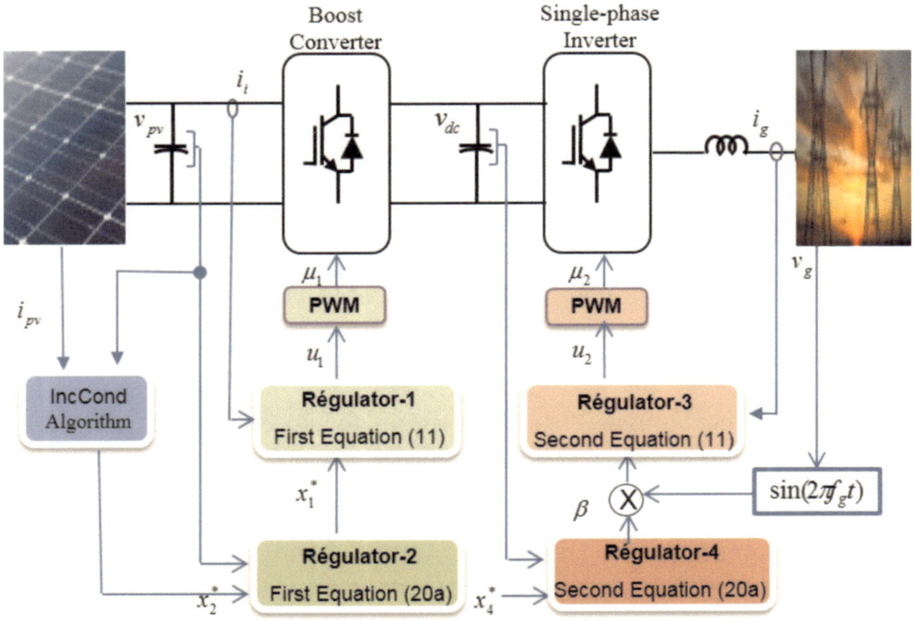

Figure 2.
Schematic diagram of the proposed controller for single phase grid-connected PV system.

$$\Omega_i = D_i'\left(x_i^*, x_i\right) - \dot{x}_i \quad (i = 1, 3) \tag{14}$$

Therefore, the control problem $\lim_{t \to \infty} e_i(t) = 0\,(i = 1, 3)$ corresponds to the insensitivity condition defined by

$$\Omega_i = 0 \quad (i = 1, 3) \tag{15}$$

Doing so, the behaviors of $\dot{x}_1$ and $\dot{x}_3$ with prescribed dynamics of expression (13) will be fulfilled. Replacing in Eq. (14) $\dot{x}_1$ and $\dot{x}_3$ by their expressions (7), (9), and since the requirement (15), one gets

$$D' - M_2 - M_1 U = 0 \tag{16}$$

Hence, there is an isolated root called the inverse dynamic solutions for U given by

$$U^{id} = \begin{pmatrix} u_1^{id} \\ u_2^{id} \end{pmatrix} = \begin{pmatrix} \dfrac{L_i}{x_4}\left(\dfrac{e_1}{T_1} - \left(-\dfrac{r_i}{L_i}x_1 + \dfrac{1}{L_i}x_2 - \dfrac{1}{L_i}x_4\right)\right) \\ \dfrac{L_g}{x_4}\left(\dfrac{e_3}{T_3} + \dfrac{x_4 + v_g + r_g x_3}{L_g}\right) \end{pmatrix} \tag{17}$$

The control variable, namely U, has emerged in Eq. (16). At this point, an appropriate control law with the first derivative in feedback has to be found, so that the (e_1, e_3)-systems are made asymptotically stable. As the objective is to drive the error to zero, it is natural to choose the Lyapunov functions candidate

$$\begin{pmatrix} V_3 \\ V_4 \end{pmatrix} = \frac{1}{2}\begin{pmatrix} (\Omega_1(u_1))^2 \\ (\Omega_3(u_2))^2 \end{pmatrix} \tag{18}$$

From expressions (14) and (16), it can be easily checked that the following time-derivatives

$$\frac{d}{dt}\begin{pmatrix} V_3 \\ V_4 \end{pmatrix} = \begin{pmatrix} \frac{\partial V_3}{\partial u_1}\frac{du_1}{dt} \\ \frac{\partial V_4}{\partial u_2}\frac{du_2}{dt} \end{pmatrix} = \begin{pmatrix} \frac{\Omega_1 \partial\Omega_1}{\partial u_1}\frac{du_1}{dt} \\ \frac{\Omega_3 \partial\Omega_3}{\partial u_2}\frac{du_2}{dt} \end{pmatrix} = -\begin{pmatrix} \frac{\Omega_1 x_4}{L_i}\frac{du_1}{dt} \\ \frac{\Omega_3 x_4}{L_g}\frac{du_2}{dt} \end{pmatrix} \tag{19}$$

are made negative-definite using the following control laws:

$$\frac{d}{dt}\begin{pmatrix} u_1 \\ u_2 \end{pmatrix} = \begin{pmatrix} \frac{k_1}{\varepsilon_1\varepsilon_2}\Omega_1 \\ \frac{k_3}{\varepsilon_1\varepsilon_2}\Omega_3 \end{pmatrix} \tag{20}$$

At this point, k_1 and k_3 are any design parameters, ε_1 and ε_2 are small positive parameters. In view of $M_1 > 0$, it is easily seem from expressions (11) and (14) that

$$\frac{d}{dt}\begin{pmatrix} V_3 \\ V_4 \end{pmatrix} = -\begin{pmatrix} \frac{k_1}{\varepsilon_1\varepsilon_2}\frac{x_4\Omega_1^2}{L_i} \\ \frac{k_3}{\varepsilon_1\varepsilon_2}\frac{x_4\Omega_3^2}{L_g} \end{pmatrix} < 0 \quad (\Omega_i \neq 0) \tag{21}$$

for a positive values of $k_i > 0$ $(i = 1, 3)$. Therefore, from expressions (14) and (20), the discussed nonlinear control laws are formulated as follows

$$\varepsilon_1\varepsilon_2\frac{d}{dt}\begin{pmatrix} u_1 \\ u_2 \end{pmatrix} = \begin{pmatrix} k_1\left(\frac{e_1}{T_1} - \frac{dx_1}{dt}\right) \\ k_3\left(\frac{e_3}{T_3} - \frac{dx_3}{dt}\right) \end{pmatrix} \tag{22}$$

3.1.2 Singular perturbation system of the inner current loops

Consider the closed-loop system of inner loops composed of the Eqs. (7)–(10) and the control laws (22), which can be rewritten in the following form

$$\varepsilon_1\varepsilon_2\frac{dZ}{dt} = h(X, Z, t) \tag{23}$$

$$\frac{dX}{dt} = f'(X, Z, t) \tag{24}$$

with $Z = (z_1\ z_2)^T = (u_1 u_2)^T$, $X = (x_1 x_2 x_3 x_4)^T$,

$$h(X,Z,t) = \begin{pmatrix} k_1\left(\frac{e_1}{T_1} - \frac{x_4}{L_i}z_1 + \frac{r_i}{L_i}x_1 - \frac{x_2}{L_i} + \frac{x_4}{L_i}\right) \\ k_3\left(\frac{e_3}{T_3} - \frac{x_4}{L_g}z_2 + \frac{v_g}{L_g} + \frac{r_g}{L_g}x_3\right) \end{pmatrix}, f'(X,Z,t) = \begin{pmatrix} z_1\frac{x_4}{L_i} - \frac{r_i}{L_i}x_1 + \frac{x_2}{L_i} - \frac{x_4}{L_i} \\ -\frac{x_1}{C_{pv}} + \frac{i_{pv}}{C_{pv}} \\ -\frac{v_g}{L_g} - \frac{r_g}{L_g}x_3 + \frac{x_4}{L_g}z_2 \\ \frac{x_1}{C_{dc}} - \frac{x_1}{C_{dc}}z_1 - \frac{x_3}{C_{dc}}z_2 \end{pmatrix}.$$

Now, we go to the fast time $\tau_1 = t/\varepsilon_1\varepsilon_2$. Then for $\varepsilon_1 = 0$, the ultra-fast dynamic subsystem (UFDS) is given by

$$\frac{dZ}{d\tau_1} = h(X, Z, t) \tag{25}$$

$$\frac{dX}{d\tau_1} = 0 \tag{26}$$

After the rapid decay of transients in expression (25), the steady state (more precisely, quasi-steady state) tends toward an equilibrium $Z^e = \tilde{h}(X)$. The manifold defined by Z^e is called the slow manifold, which is given by

$$Z^e = \begin{pmatrix} z_1^e \\ z_2^e \end{pmatrix} = \begin{pmatrix} \dfrac{L_i}{x_4}\left(\dfrac{e_1}{T_1} + \dfrac{x_4 - x_2 + r_i x_1}{L_i}\right) \\ \dfrac{L_g}{x_4}\left(\dfrac{e_3}{T_3} + \dfrac{v_g + r_g x_3}{L_g}\right) \end{pmatrix} \tag{27}$$

Remark 2.

i. During the fast transient in expression (25), the variables X are treated as the frozen parameters.

ii. The equilibrium point given by expression (27) involves a division by the DC link voltage x_4, from a practical point of view this division is not a problem because the DC link voltage remains all the time positive for the power converter to work correctly.

By substituting of this equilibrium Z^e into Eq. (24), the slow dynamic subsystem (SDS) of inner loops takes place on the slow manifold, according to the equation

$$\frac{dX}{dt} = \begin{pmatrix} \dfrac{e_1}{T_1} \\ -\dfrac{x_1}{C_{pv}} + \dfrac{i_{pv}}{C_{pv}} \\ \dfrac{e_3}{T_3} \\ -\dfrac{L_o x_3}{C_{dc} x_4}\left(\dfrac{e_3}{T_3} + \dfrac{v_g + r_g x_3}{L_g}\right) - \dfrac{L_i x_1}{C_{dc} x_4}\left(\dfrac{e_1}{T_1} + \dfrac{r_i x_1 - x_2}{L_i}\right) \end{pmatrix} \tag{28}$$

Proposition 1. Consider the closed-loop system composed of Eqs. (23) and (24). For $\varepsilon_1 \to 0$, the system takes the singular perturbation form where the UFDS is defined by equation (25), while the SDS of inner loop is defined by equation (28). Under the considerations given by Remark 2, one has the following properties

i. If the design parameters k_i $(i = 1, 3)$ are positives, the UFDS (25) will be exponentially stable, and Z converge exponentially fast to Z^e.

ii. The behaviors of x_i $(i = 1, 3)$ are prescribed by the stable reference equations of the form $dx_i/dt = \left(x_i^* - x_i\right)/T_i$. Then, the requirements $\lim\limits_{t\to\infty} e_i = \lim\limits_{t\to\infty}\left(x_i^* - x_i\right) = 0$ are maintained.

3.2 MPPT and DC bus voltage regulation objective

The second step consists in completing the inner control loops by outer control loops for PV voltage (Regulator 2) and DC-link voltage (Regulator 4). The aim is now to enforce the photovoltaic voltage x_2 and the DC-link voltage x_4 to track, respectively, the optimal point $x_2^* = V'_M$ and a given reference voltage $x_4^* = V'_{dc}$, such that tuning laws for the ratio β and x_1^* must be designed. According to the three-time-scale design methodology that is employed in this work, the general formulation of the three-time-scale singular perturbed systems requires the system to possess three different time scales. To this end, the voltage loops will be slow compared to the transients of the current loops. Therefore, the design parameters for voltage loops, in particular (ε_2, T_2, and T_4) must satisfy: $0 < \varepsilon_1\varepsilon_2 \ll \varepsilon_2 \ll 1$, and $(T_1, T_3) < (T_2, T_4)$. In addition, the steady states for the current x_1 and x_3 yield, respectively, x_1^* and x_3^*. Therefore, the SDS of inner loop given by equation (28) will be reduced to

$$\frac{dX_r}{dt} = \frac{d}{dt}\begin{pmatrix} x_2 \\ x_4 \end{pmatrix} = \begin{pmatrix} \dfrac{i_{pv}}{C_{pv}} - \dfrac{x_1^*}{C_{pv}} \\ \dfrac{x_2x_1^* - v_gx_3^* - r_i\left(x_1^*\right)^2 - r_g\left(x_3^*\right)^2}{C_{dc}x_4} \end{pmatrix} \tag{29}$$

Now, the first step is to establish the relation between the ratio β (which acts as the control input of the outer loop) and the DC-link voltage x_4 (representing the output of the outer loop).

3.2.1 Relation between (β) *and* (x_4)*, and the control law*

The relation between the ratio β and the DC-link voltage x_4 is the subject of the following proposition.

Proposition 2. We consider the second equation of expression (29) and the power factor correction requirement defined by $x_3^* = \beta \sin\left(\omega_g t\right)$.

i. The relation between x_4 and β is described in the following results

$$\frac{dx_4}{dt} = \frac{x_2x_1^* - r_i\left(x_1^*\right)^2}{C_{dc}x_4} - \frac{v_g^2\beta}{C_{dc}x_4E_g} - \frac{r_gv_g^2\beta^2}{C_{dc}x_4E_g^2} \tag{30}$$

ii. Therefore, the squared-voltage $x_{r,4} = x_4^2$ varies, in response to the tuning ratio β, according to the following first-order time-varying nonlinear equation:

$$\frac{dx_{r,4}}{dt} = 2\frac{x_2x_1^* - r_i\left(x_1^*\right)^2}{C_{dc}} - \frac{2v_g^2\beta}{C_{dc}E_g} - \frac{2r_gv_g^2\beta^2}{C_{dc}E_g^2} \tag{31}$$

The second step is to establish control laws for the outer loops, in which $Y = \left(y_1y_2\right)^T = \left(x_1^*\beta\right)^T$ represent the new control inputs, while $X_r = \left(x_{r,2}x_{r,4}\right)^T = \left(x_2x_4^2\right)^T$ represent the new output variables. To this end, introduce the following tracking errors

$$e_j = x_j^* - x_{r,j} \quad (j = 2, 4) \tag{32}$$

Then, let the desired behavior of X_r be assigned by

$$\frac{dx_{r,j}}{dt} = \frac{e_j}{T_j} \triangleq D_j\left(x_j^*, x_j\right) \quad (j = 2, 4) \tag{33}$$

The error of the desired dynamic realization it follows

$$\Omega_j = D_j\left(x_{r,j}^*, x_{r,j}\right) - \dot{x}_{r,j} \quad (j = 2, 4) \tag{34}$$

Then, the insensitivity condition is given by

$$\Omega_j = 0 \tag{35}$$

Similar to the previous subsection and bearing in mind the fact that β and their first derivative must be available, and in order to meet the requirement (35), we should apply the control law given by the following structure

$$\varepsilon_2 \frac{d}{dt}\begin{pmatrix} y_1 \\ y_2 \end{pmatrix} = \begin{pmatrix} k_2(e_2/T_2 - dx_{2,r}/dt) \\ k_4(e_4/T_4 - dx_{r,4}/dt) \end{pmatrix} \tag{36}$$

3.2.2 Singular perturbation system of the outer voltage loops

Combining expressions (31) and (36), and the first equation of expression (29), one obtains

$$\varepsilon_2 \frac{dY}{dt} = g(X_r, Y, t) = \begin{pmatrix} k_2\left(\dfrac{e_2}{T_2} + \dfrac{y_1 - i_{pv}}{C_{pv}}\right) \\ k_4\left(\dfrac{e_4}{T_4} - f_2''\right) \end{pmatrix} \tag{37}$$

$$\frac{dX_r}{dt} = f''(X_r, Y, t) = \begin{pmatrix} \dfrac{i_{pv}}{C_{pv}} - \dfrac{1}{C_{pv}} y_1 \\ f_2'' \end{pmatrix} \tag{38}$$

with $f_2'' = 2\left(\frac{x_{r,2} y_1 - r_i y_1^2}{C_{dc}} - \frac{E_g v_g^2 y_2 + r_g v_g^2 y_2^2}{C_{dc} E_g^2}\right)$. The fast dynamic subsystem (FDS) is obtained by transforming the slow time-scale t to the fast time-scale $\tau_2 = t/\varepsilon_2$, then, by setting $\varepsilon_2 = 0$

$$\frac{dY}{d\tau_2} = g(X_r, Y, t) \tag{39}$$

$$\frac{dX_r}{d\tau_2} = 0 \tag{40}$$

Notice that expression (39) has an isolated equilibrium at $Y^e = \left(y_1^e y_2^e\right)$, which will be determined (in the mean) in Appendix. As the FDS (39) is nonlinear, the stability properties of its equilibrium can be checked through the analysis of the Jacobian matrix of the linearized version defined as follows

$$A_F = \begin{bmatrix} \dfrac{k_2}{C_{pv}} & 0 \\ k_4 A_F' & k_4 A_F'' \end{bmatrix} \tag{41}$$

DOI: http://dx.doi.org/10.5772/intechopen.89434

with $A'_F = \frac{4r_i y_1^e}{C_{dc}} - \frac{2x_{r,2}}{C_{dc}}$, $A''_F = \frac{2v_g^2}{C_{dc}E_g} + \frac{4r_g v_g^2 y_2^e}{C_{dc}E_g^2}$. Taking into account that y_2^e is positive, we conclude that all eigenvalues of A_F satisfy $\text{Re}\,(\lambda_{1,2}) < 0$ for the negative values of k_2 and k_4. Therefore, A_F is Hurwitz matrix. By substituting of the equilibrium Y^e into expression (38), the reduced SDS of outer loops takes place on the slow manifold, according to the equation

$$\frac{dX_r}{dt} = \begin{pmatrix} e_2/T_2 \\ e_4/T_4 \end{pmatrix} \tag{42}$$

Proposition 3. Consider the system closed loop composed of expressions (37) and (38). For $\varepsilon_2 \to 0$, this system takes the singular perturbation form, where the FDS is given by expression (39) and the reduced SDS of outer loop is given by expression (42). One has the following properties

i. If the design parameters k_2 and k_4 are negative, the FDS (39), will be exponentially stable and Y converge exponentially fast to Y^e.

ii. The behaviors of $x_{r,j}$ $(j = 2, 4)$ are prescribed by the stable reference equations of the form $dx_{r,j}/dt = \left(x_j^* - x_{r,j}\right)/T_j$. Then, the requirements $\lim_{t\to\infty} e_j = \lim_{t\to\infty}\left(x_j^* - x_{r,j}\right) = 0$ $(j = 2, 4)$ are maintained.

4. Control system analysis

The objective of the global stability of closed-loop system can be analyzed in the following theorem. It is shown that the control objectives are achieved (in the mean) with an accuracy that depends on the network frequency ω_g and the small parameters $\varepsilon_i^{\otimes}$ $(i = 1, 2)$.

Theorem. Consider the overall control system composed of the Pv panel, boost DC-DC converter and DC-AC inverter, described by the model (7)–(10), in closed loop with the multi-cascade multi-loop composed of:

- The inner regulators (23), where $(\varepsilon_1\varepsilon_2, k_1, k_3, T_1, T_3)$ are the design parameters;
- The outer regulators (37), where $(\varepsilon_2, k_2, k_4, T_2, T_4)$ are the design parameters.

Then, one has the following property

i. The augmented state vector $Z = (u_1\ u_2)^T = (z_1\ z_2)^T$, $Y = \left(x_1^*\beta\right)^T = \left(y_1 y_2\right)^T$, and $X' = \left(x'_1 x'_2 x'_3 x'_4\right)^T = \left(x_1 x_2 x_3 x_4^2\right)^T$ undergoes the following state equations

$$\varepsilon_1\varepsilon_2\dot{Z}(t) = h(X', Y, Z, t) \tag{43}$$

$$\varepsilon_2\dot{Y}(t) = g(X', Y, Z, t) \tag{44}$$

$$\dot{X}'(t) = f(X', Y, Z, t) \tag{45}$$

with

$$h = \begin{pmatrix} k_1\left(\frac{e_1}{T_1} - \frac{z_1\sqrt{x_4'}}{L_i} + \frac{r_i x_1' - x_2' + \sqrt{x_4'}}{L_i}\right) \\ k_3\left(\frac{e_3}{T_3} - \frac{\sqrt{x_4'}}{L_g} z_2 + \frac{v_g + r_g x_3'}{L_g}\right) \end{pmatrix}, f = \begin{pmatrix} -\frac{r_i}{L_i}x_1' + \frac{x_2'}{L_i} - \frac{\sqrt{x_4'}}{L_i} + z_1\frac{\sqrt{x_4'}}{L_i} \\ -\frac{y_1}{C_{pv}} + \frac{i_{pv}}{C_{pv}} \\ \frac{\sqrt{x_4'}}{L_o} z_2 - \frac{v_g}{L_g} - \frac{r_o}{L_g}x_3' \\ 2\frac{x_2' y_1 - r_i y_1^2}{C_{dc}} - \frac{2v_g^2 y_2}{C_{dc}E_g} - \frac{2r_g v_g^2 y_2^2}{C_{dc}E_g^2} \end{pmatrix},$$

$$g = \begin{pmatrix} k_2\left(\frac{e_2}{T_2} + \frac{y_1}{C_{pv}} - \frac{i_{pv}}{C_{pv}}\right) \\ k_4\left(\frac{e_4}{T_4} + \frac{2v_g^2 y_2}{C_{dc}E_g} + \frac{2r_g v_g^2 y_2^2}{C_{dc}E_g^2} - 2\frac{x_2' y_1 - r_i y_1^2}{C_{dc}}\right) \end{pmatrix}.$$

ii. Let the control design parameters be selected, such that the following inequalities hold $T_1 < T_2$, $T_3 < T_4$, $k_1 > 0$, $k_2 < 0$, $k_3 > 0$, $k_4 < 0$, and $0 < \varepsilon_1\varepsilon_2 \ll \varepsilon_2 \ll 1$.

Then, there exist positive constants $\rho^{\otimes}$, $\varepsilon_n^{\otimes}$ $(n = 1, 2)$, such that for and $0 < \rho < \rho^{\otimes}$ $0 < \varepsilon_n < \varepsilon_n^{\otimes} \ll 1$ $(n = 1, 2)$ the system (43)–(45) has an asymptotically stable $(2\pi/\omega_g)$ periodic solution $X(t, \varepsilon_n, \rho)$, $Y(t, \varepsilon_n, \rho)$, $Z(t, \varepsilon_n, \rho)$ that continuously depends on ε_n $(n = 1, 2)$ and $\rho = 1/\omega_g$. Furthermore, when ($\varepsilon_n = 0$ and $\rho = 0$), one has

$$\lim_{\varepsilon_n \to 0, \rho \to 0} X(t, \varepsilon_n, \rho) = X_0^*, \quad \lim_{\varepsilon_n \to 0, \rho \to 0} Y(t, \varepsilon_n, \rho) = Y_0^*, \quad \lim_{\varepsilon_n \to 0, \rho \to 0} Z(t, \varepsilon_n, \rho) = Z_0^*, \tag{46}$$

where $X_0^* = \left(x_{1,0}^*\ x_{2,0}^*\ x_{3,0}^*\ x_{4,0}^*\right)^T$, $Y_0^* = \left(y_{1,0}^* y_{2,0}^*\right)^T$, and $Z_0^* = \left(z_{1,0}^*\ z_{2,0}^*\right)^T$ with $x_{2,0}^* = V_M'$, $x_{4,0}^* = V_{dc}'$, $z_{1,0}^* = 1 - \frac{x_{2,0}^*}{\sqrt{x_{4,0}^*}} + \frac{r_i x_{1,0}^*}{\sqrt{x_{4,0}^*}}$, $z_{2,0}^* = \frac{r_o x_{3,0}^*}{\sqrt{x_{4,0}^*}}$, $y_{1,0}^* = i_{pv,0}$,

$y_{2,0}^* = \frac{-E_g + 2\sqrt{(E_g/2)^2 + 2r_o\left(x_{2,0}^* i_{pv,0} - r_i\left(i_{pv,0}\right)^2\right)}}{2r_o}$.

See Appendix for the proof.

5. Simulation and discussion of results

The experimental setup is described by **Figure 2** and the nonlinear controller, developed in Section 3, including the control laws (23) and (37), will now be evaluated by simulation in MATLAB/SIMULINK platform using the electromechanical characteristics of **Table 1**.

The numerical values used for the design parameters are chosen as follows: $\varepsilon_1 = 5.81 \times 10^{-2}$, $\varepsilon_2 = 7.21 \times 10^{-5}$, $T_1 = 2.1 \times 10^{-4}$, $T_2 = 9.3 \times 10^{-3}$, $T_3 = 9.51 \times 10^{-5}$, $T_4 = 1.43 \times 10^{-2}$, $k_1 = 7.1 \times 10^{-7}$, $k_2 = -2.56 \times 10^{-6}$, $k_3 = 2.3 \times 10^{-6}$, $K_4 = 8.1 \times 10^{-8}$. These values have proved to be suitable based on several trials respecting the singular perturbation technique. In this simulation, we consider the KC200GT type of PV array module with $N_s = 54$ and $N_p = 6$.

The performances of the proposed controller are illustrated by **Figures 3–5**. **Figure 3** shows that the DC-link voltage x_4 is well regulated and quickly settles

Parameters	Symbol	Values
Network	E/f	$220\sqrt{2}V/50Hz$
Boost	C_{pv} L_i r_i	0.3mF 6mH 20mΩ
L-filter	L_o r_o	5mH 50mΩ
PWM switching frequency	f_{PWM}	14kHz
DC capacitance	C_{dc}	7mF

Table 1.
PV system and single-phase grid characteristics.

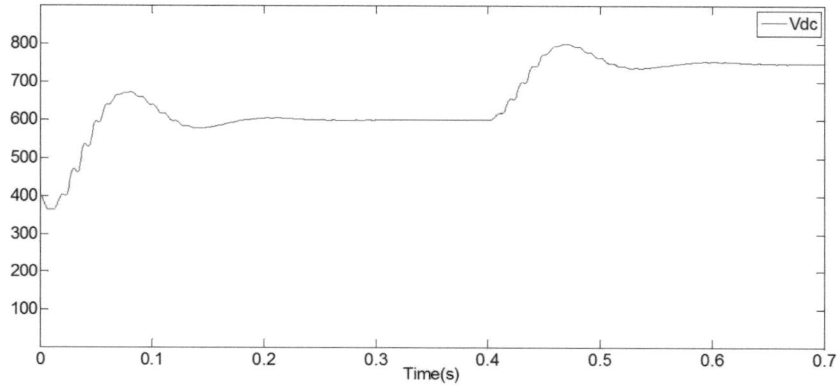

Figure 3.
DC-link voltage.

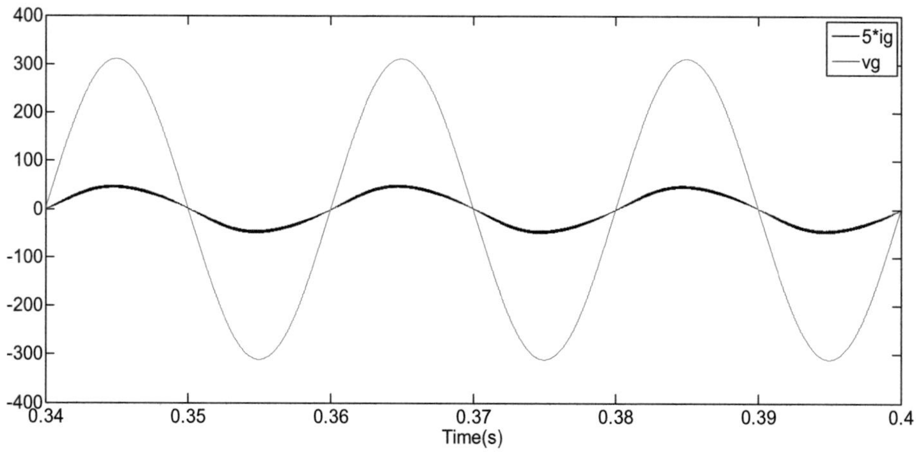

Figure 4.
Power factor correction checking.

down after each change in the signal reference (stepping from $600V$ to $750V$ at $t = 0.4s$). The wave frame of the output current x_3 is showed in **Figure 4**. The current is sinusoidal and in phase with the network voltage complying with the PFC requirement. This is further demonstrated by **Figure 5**, which shows that the ratio β takes a constant value after transient periods following the changes in reference signals.

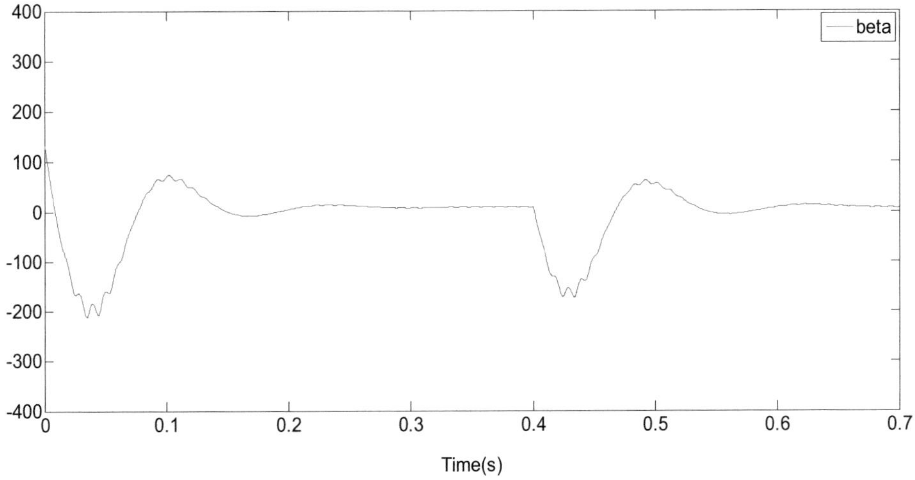

Figure 5.
Tuning parameter β.

5.1 Radiation variation effect

Figure 6a–c illustrates the resulting closed-loop control performances in presence of radiation changes. Specifically, the radiation takes a low, medium, and high value (equal to 800, 1000 and step to 600 W/m^2 at times 0, 0.4, and 0.7 s, respectively), meanwhile the temperature is kept constant, equal to 298.17K (i.e. 25$C°$). **Figure 6a** shows that the captured PV voltage varies between 182, 212, and 138 V. These values correspond to the maximum points. **Figure 6b** shows that the DC-link voltage regulation is recovered after a short transient period following each change of the irradiation. **Figure 6c** shows that the current amplitude changes whenever the radiation varies. It is seen that the output current x_3 and the grid voltage v_g are actually sinusoidal and in phase. Hence, the converter connection to the supply network is done with a unitary power factor.

5.2 Temperature variation effect

The perfect MPPT is illustrated by **Figure 7a**. Here, the temperature steps from 298.15 to 318.15 K, then to 308.15 K while the radiation λ is kept constant equal to 1000W/m^2. **Figure 7b** shows that the DC-link voltage x_4 is tightly regulated: it quickly settles down after each change in the temperature. **Figure 7c** illustrates the current amplitude changes whenever the temperature varies. The current remains (almost) sinusoidal and in phase with the network voltage complying with the PFC requirement.

6. Conclusion

In this chapter, an advanced controller is developed for PV grid-connected system. The latter is described by fourth-order nonlinear averaged model. The multi-loops nonlinear controller has been designed and developed using three-time singular perturbation technique and averaging theory.

Using the theoretical analysis (via three-time-scale singular perturbation technique and averaging theory) and simulation, it is proved that the controller

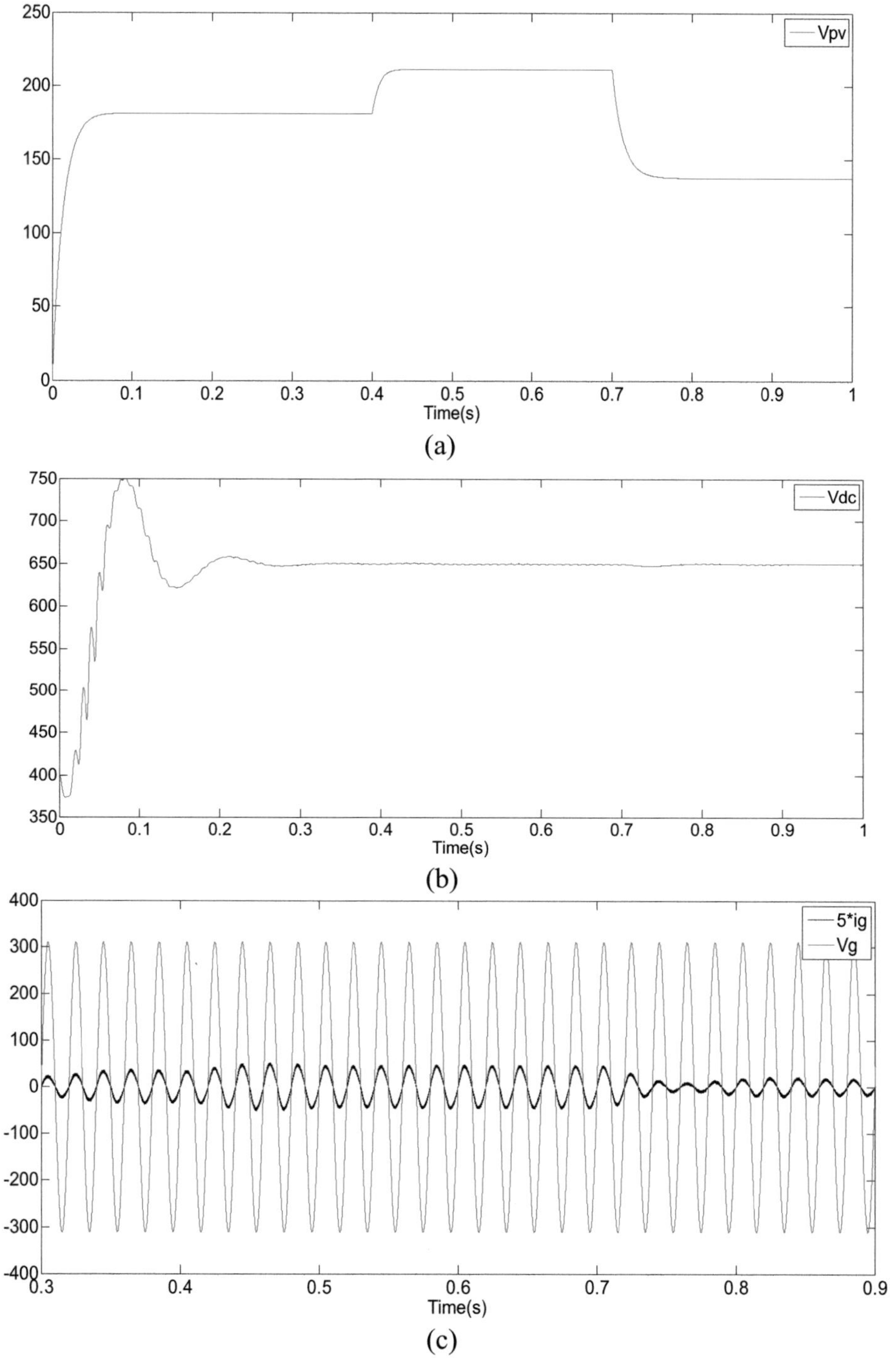

Figure 6.
Radiation variation effect: (a) Photovoltaic voltage. (b) DC-link voltage. (c) PFC checking.

does meet the performances for which it was designed, namely: (i) Maximum power point tracking of PV array; (ii) tight regulation of the DC bus voltage; (iii) perfect power factor in the grid; and (iv) global asymptotic stability of the all system.

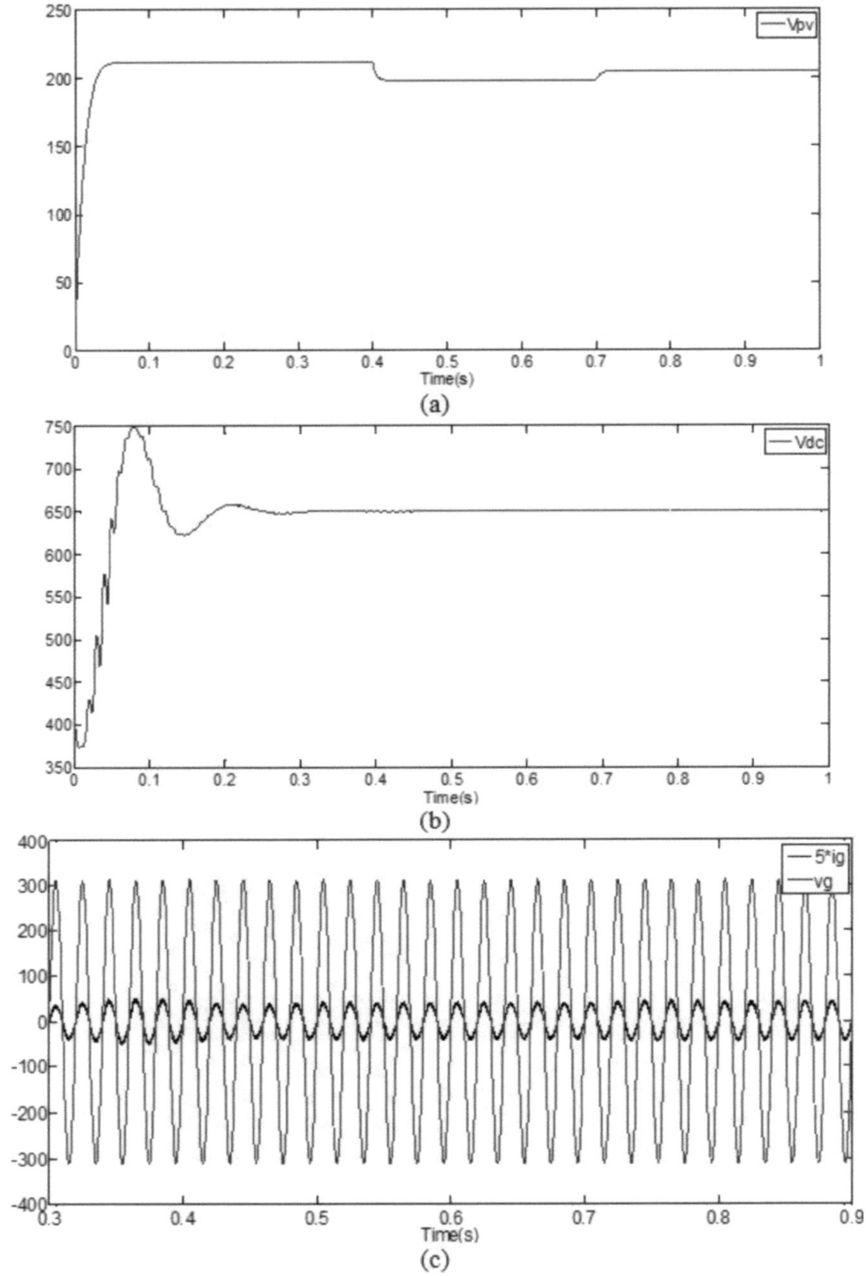

Figure 7.
Temperature variation effect: (a) Photovoltaic voltage. (b) DC-link voltage. (c) PFC checking.

Several simulation results have been made that illustrate the high performances of the proposed controller in ideal operating conditions (in the presence of meteorological constant) and its robustness against radiation and temperature change.

Appendix (Proof)

Part 1: Eqs. (43)–(45) are immediately obtained from expressions (37), (38), (23), and the first and the third equation of expression (24).

Part 2: The stability of the time-varying system (43)–(45) will now be performed in two steps using the averaging theory (e.g. Chapter 10 in Refs. [19, 25]) and the singular perturbation theory (e.g., Chapter 11 in Refs. [19, 23, 24]). The next step consists in using the averaging theory. To this end, let us introduce the time-scale change $\bar{t} = \omega_g t$. Using this time, one gets $\varepsilon_1\varepsilon_2\dot{\overline{Z}}(\bar{t}) = \rho\bar{h}\left(\overline{X}',\overline{Z},\bar{t}\right)$, $\varepsilon_2\dot{\overline{Y}}(\bar{t}) = \rho\bar{g}\left(\overline{X}',\overline{Y},\bar{t}\right)$, and $\dot{\overline{X}}'(\bar{t}) = \rho\bar{f}\left(\overline{X}',\overline{Y},\overline{Z},\bar{t}\right)$ with $\rho = 1/\omega_g$. In view of expressions (43)–(45), it is seem that the functions $\bar{h}\left(\overline{X}',\overline{Z},\bar{t}\right)$, $\bar{g}\left(\overline{X}',\overline{Y},\bar{t}\right)$, and $\bar{f}\left(\overline{X}',\overline{Y},\overline{Z},\bar{t}\right)$ as functions of $\bar{t}$ are periodic with period–2π, let us introduce the averaged functions:

$$\varepsilon_1\varepsilon_2\dot{Z}_0(\bar{t}) = \rho\lim_{t\to\infty}\frac{1}{2\pi}\int\bar{h}\left(\overline{X}',\overline{Z},\bar{t}\right)d\bar{t} = \rho h_0(X_0,Z_0) \tag{47}$$

$$\varepsilon_2\dot{Y}_0(\bar{t}) = \rho\lim_{\rho\to 0}\frac{1}{2\pi}\int\bar{g}\left(\overline{X}',\overline{Y},\bar{t}\right)d\bar{t} = \rho g_0(X_0,Y_0) \tag{48}$$

$$\dot{X}_0(\bar{t}) = \rho\lim_{\rho\to 0}\frac{1}{2\pi}\int\bar{f}\left(\overline{X}',\overline{Y},\overline{Z},\bar{t}\right)d\bar{t} = \rho f_0(X_0,Y_0,Z_0) \tag{49}$$

Since the systems here studied present equilibrium different from zero and in order to satisfy this requirement, a change of variables is introduced such that defines the new system in terms of its error dynamics. Therefore, the error dynamics are defined by introducing: $\tilde{Z}_0 = (\tilde{z}_{1,0}\tilde{z}_{2,0})^T = \left(\left(z_{1,0}-z^*_{1,0}\right)\left(z_{2,0}-z^*_{2,0}\right)\right)^T$, $\tilde{Y}_0 = \left(\tilde{y}_{1,0}\quad\tilde{y}_{2,0}\right)^T = \left(\left(y_{1,0}-y^*_{1,0}\right)\left(y_{2,0}-y^*_{2,0}\right)\right)^T$, and $\tilde{X}_0 = (\tilde{x}_{1,0}\quad\tilde{x}_{2,0}\quad\tilde{x}_{3,0}\quad\tilde{x}_{4,0}) = \left(\left(x_{1,0}-x^*_{1,0}\right)\left(x_{2,0}-x^*_{2,0}\right)\left(x_{3,0}-x^*_{3,0}\right)\left(x_{4,0}-x^*_{4,0}\right)\right)$, where the constant $x^*_{1,0}$, $x^*_{2,0}$, $x^*_{3,0}$, $x^*_{4,0}$, $y^*_{1,0}$, $y^*_{2,0}$, $z^*_{1,0}$, and $z^*_{2,0}$ represent the desired average values of the state variables. Then, since expressions (43)–(45) and according to expressions (47)–(49), the system can be rewritten into its error-dynamics formulation thus defining the closed-loop error dynamics as:

$$\varepsilon_1\varepsilon_2\dot{\tilde{Z}}_0 = \rho\tilde{h}_0\left(\tilde{X}_0,\tilde{Z}_0\right) = \rho\begin{pmatrix} k_1\left(-\dfrac{\tilde{x}_{1,0}}{T_1}-\tilde{f}_{1,0}\left(\tilde{X}_0,\tilde{Z}_0\right)\right) \\ k_3\left(-\dfrac{\tilde{x}_{3,0}}{T_3}-\tilde{f}_{3,0}\left(\tilde{X}_0,\tilde{Z}_0\right)\right)\end{pmatrix} \tag{50}$$

$$\varepsilon_2\dot{\tilde{Y}}_0 = \rho\tilde{g}_0\left(\tilde{X}_0,\tilde{Y}_0\right) = \rho\begin{pmatrix} k_2\left(-\dfrac{\tilde{x}_{2,0}}{T_2}-\tilde{f}_{2,0}\left(\tilde{X}_0,\tilde{Y}_0\right)\right) \\ k_4\left(-\dfrac{\tilde{x}_{4,0}}{T_4}-\tilde{f}_{4,0}\left(\tilde{X}_0,\tilde{Y}_0\right)\right)\end{pmatrix} \tag{51}$$

$$\dot{\tilde{X}}_0 = \rho\tilde{f}_0\left(\tilde{X}_0,\tilde{Y}_0,\tilde{Z}_0\right) = \rho\begin{pmatrix}\tilde{f}_{1,0}\left(\tilde{X}_0,\tilde{Z}_0\right) \\ \tilde{f}_{2,0}\left(\tilde{X}_0,\tilde{Y}_0\right) \\ \tilde{f}_{3,0}\left(\tilde{X}_0,\tilde{Z}_0\right) \\ \tilde{f}_{4,0}\left(\tilde{X}_0,\tilde{Y}_0\right)\end{pmatrix} \tag{52}$$

where

$$\tilde{f}_{1,0}\left(\tilde{X}_0,\tilde{Z}_0\right) = \frac{\left(\tilde{x}_{2,0}+x^*_{2,0}\right)-\sqrt{\left(\tilde{x}_{4,0}+x^*_{4,0}\right)}}{L_i} - \frac{r_i\left(\tilde{x}_{1,0}+x^*_{1,0}\right)}{L_i} + \frac{\left(\tilde{z}_{1,0}+z^*_{1,0}\right)\sqrt{\left(\tilde{x}_{4,0}+x^*_{4,0}\right)}}{L_i} \tag{53}$$

$$\tilde{f}_{2,0}(\tilde{X}_0, \tilde{Y}_0) = \frac{i_{pv,0}}{C_{pv}} - \frac{\left(\tilde{y}_{1,0} + y^*_{1,0}\right)}{C_{pv}} \tag{54}$$

$$\tilde{f}_{3,0}(\tilde{X}_0, \tilde{Z}_0) = -\frac{r_g\left(\tilde{x}_{3,0} + x^*_{3,0}\right)}{L_g} + \frac{\sqrt{\left(\tilde{x}_{4,0} + x^*_{4,0}\right)}}{L_g}\left(\tilde{z}_{2,0} + z^*_{2,0}\right) \tag{55}$$

$$\begin{aligned}\tilde{f}_{4,0}(\tilde{X}_0, \tilde{Y}_0) = &-\frac{E_g\left(\tilde{y}_{2,0} + y^*_{2,0}\right) + r_g\left(\tilde{y}_{2,0} + y^*_{2,0}\right)^2}{C_{dc}} \\ &+ 2\frac{\left(\tilde{x}_{2,0} + x^*_{2,0}\right)\left(\tilde{y}_{1,0} + y^*_{1,0}\right) - r_i\left(\left(\tilde{y}_{1,0} + y^*_{1,0}\right)\right)^2}{C_{dc}}\end{aligned} \tag{56}$$

In order to get stability results regarding the system of interest (43)–(45), it is sufficient (thanks to averaging theory) to analyze the averaged system (50)–(52). Now, the asymptotic stability of the resulting three-time-scale photovoltaic single phase grid system (50)–(52) is discussed, which is based on the sequential (double) time-scale analysis similar to the one presented in Refs. [23, 24], it is an extension of the two-time-scale analysis presented [19].

The use of theory of the three-time-scale singular perturbations for the stability analysis is based on the idea that, for $0 < \varepsilon_1\varepsilon_2 \ll \varepsilon_2 \ll 1$, the trajectories in $\tilde{X}_0$, $\tilde{Y}_0$, and $\tilde{Z}_0$ of the system (50)–(52) can be approximated by three models: the slow dynamic subsystem (SDS) of full system, the fast dynamic subsystem (FDS), and the ultra-fast dynamic subsystem (UFDS). We can thus find Lyapunov functions for each one of the singularly perturbed subsystems.

For the UFDS, it is necessary to ensure that the dynamic of expression (50) does not to shift from the quasi-steady-state equilibrium $\tilde{Z}_0 = \tilde{h}'_0(\tilde{X}_0)$. Then, the associated Lyapunov function candidate is obtained by introducing a change of variables $\hat{Z}_0 = \tilde{Z}_0 - \tilde{h}'_0(\tilde{X}_0)$, so that its equilibrium is centered at zero. By letting $\tau_1 = \bar{t}/\varepsilon_1\varepsilon_2$, the UFSD in function of $\hat{Z}_0$ is defined as follows

$$\frac{d\tilde{Z}_0}{d\tau_1} = \rho\hat{h}_0(\tilde{X}_0, \hat{Z}_0) = \rho\begin{pmatrix} -\dfrac{k_1\sqrt{\left(\tilde{x}_{4,0} + x^*_{4,0}\right)}}{L_i}\hat{z}_{1,0} \\ -\dfrac{k_3\sqrt{\left(\tilde{x}_{4,0} + x^*_{4,0}\right)}}{L_g}\hat{z}_{2,0}\end{pmatrix} \tag{57}$$

in which $\tilde{X}_0$ is treated as a fixed parameter. Thus, the associated Lyapunov function can be defined by

$$V_U = \frac{1}{2\rho}\left(P_{U1}\hat{z}^2_{1,0} + P_{U2}\hat{z}^2_{2,0}\right) \tag{58}$$

In view of $k_i > 0\,(i = 1, 3)$ and $\left(\tilde{x}_{4,0} + x^*_{4,0}\right) > 0$, it is clear that the solutions P_{U1} and P_{U2} can be chosen as

$$P_{U1} = \frac{L_i}{2k_1\sqrt{\left(\tilde{x}_{4,0} + x^*_{4,0}\right)}}q_{U1} \tag{59}$$

$$P_{U2} = \frac{L_g}{2k_3\sqrt{\left(\tilde{x}_{4,0} + x^*_{4,0}\right)}} q_{U2} \tag{60}$$

with q_{U1} and q_{U2} are positive. Similar to the ultra-fast subsystem, it is necessary to ensure that the dynamic of expression (51) does not to shift from the equilibrium $\tilde{Y}_0 = \tilde{g}'_0(\tilde{X}_0)$. Then, by introducing a change of variables $\hat{Y}_0 = \tilde{Y}_0 - \tilde{g}'_0(\tilde{X}_0)$, so that its equilibrium is centered at zero and by letting $\tau_2 = \bar{t}/\varepsilon_2$, the ultra-fast subsystem can be rewritten in the following form

$$\frac{d\tilde{Y}_0}{d\tau_2} = \rho\hat{g}_0(\tilde{X}_0, \hat{Y}_0) = \rho\begin{pmatrix} \frac{k_2}{C_{pv}}\hat{y}_{1,0} \\ \psi''(\tilde{X}_0, \hat{Y}_0) \end{pmatrix} \tag{61}$$

with $\psi'' = 2k_4\left[\frac{r_i\hat{y}^2_{1,0}}{C_{dc}} + \frac{r_g\hat{y}^2_{2,0}}{2C_{dc}} + \frac{\psi'\hat{y}_{2,0}}{C_{dc}} + \left(\frac{2r_iC_{pv}}{C_{dc}}\left(\frac{\tilde{x}_{2,0}}{T_2} + \frac{i_{pv,0}}{C_{pv}}\right) - \frac{\left(\tilde{x}_{2,0}+x^*_{2,0}\right)}{C_{dc}}\right)\hat{y}_{1,0}\right]$.

Since expression (61), it is seem that the fast subsystem is nonlinear. According to the proposition 4 (part i), it is shown that this subsystem can be made asymptotically stable by letting $k_j > 0$ (for $j = 2, 4$). Moreover, we can find a Lyapunov function, which takes the following quadratic form

$$V_F = \frac{1}{2\rho}\left(p_{F1}\hat{y}^2_{1,0} + p_{F2}\hat{y}^2_{2,0} + 2p_{F3}\hat{y}_{1,0}\hat{y}_{2,0}\right) \tag{62}$$

with the solutions to the associated Lyapunov given as

$$p_{F1} = -\frac{C_{pv}q_{F1}}{2k_2} - \frac{2\left(r_iC_{pv}\left(\frac{\tilde{x}}{2,0} + \frac{i_{pv,0}}{C_{pv}}\right) - \frac{\left(\tilde{x}_{2,0}+x^*_{2,0}\right)}{2}\right)^2 q_{F2}}{k_2\psi'\left(\frac{\psi'}{C_{pv}} + \frac{C_{dc}k_2}{2k_4C^2_{pv}}\right)} \tag{63}$$

$$p_{F2} = -\frac{C_{dc}}{4k_4\psi'}q_{F2} \tag{64}$$

$$p_{F3} = \frac{C_{dc}\left(r_iC_{pv}\left(\frac{\tilde{x}}{2,0} + \frac{i_{pv,0}}{C_{pv}}\right) - \frac{\left(\tilde{x}_{2,0}+x^*_{2,0}\right)}{2}\right)}{k_4\psi'\left(2\psi' + \frac{C_{dc}k_2}{k_4C_{pv}}\right)}q_{F2} \tag{65}$$

in which $\tilde{X}_0$ is treated as a fixed parameter. Finally, the new slow dynamic subsystem is obtained by substituting of the ultra-fast subsystem equilibrium $\tilde{Z}_0 = \tilde{h}'_0(\tilde{X}_0)$ and the fast subsystem equilibrium $\tilde{Y}_0 = \tilde{g}'_0(\tilde{X}_0)$ into expression (52)

$$\dot{\tilde{X}}_0 = \left(\dot{\tilde{x}}_{1,0} \quad \dot{\tilde{x}}_{2,0} \quad \dot{\tilde{x}}_{3,0} \quad \dot{\tilde{x}}_{4,0}\right)^T = -\rho(\tilde{x}_{1,0}/T_1 \quad \tilde{x}_{2,0}/T_2 \quad \tilde{x}_{3,0}/T_3 \quad \tilde{x}_{4,0}/T_4)^T \tag{66}$$

It is easy to define the associated Lyapunov function for SDS as follows

$$V_S(\tilde{X}_0) = \frac{1}{2\rho}\tilde{X}^T_0P_s\tilde{X}_0 = \frac{1}{2\rho}(p_{S1}x^2_{1,0} + p_{S2}x^2_{2,0} + p_{S3}x^2_{3,0} + p_{S4}x^2_{4,0}) \tag{67}$$

where $p_{S1} = T_1q_{S1}/2$, $p_{S2} = T_2q_{S2}/2$, $p_{S3} = T_3q_{S3}/2$, $p_{S4} = T_4q_{S4}/2$, and q_{S1}, q_{S2}, q_{S3}, q_{S4} are the positive constants.

Based on these Lyapunov function candidates, the double application of the standard two-time-scale stability analysis is divided in two stages: in the first stage, the stability analysis focusses on proving the stability properties of the degenerated $\sum_{SF}$-subsystem (slow-fast subsystem). The results obtained will be used in the second stage in order to prove the stability properties for the full $\sum_{SFU}$-system (slow-fast-ultra-fast subsystem).

$\sum SF$ **Stability analysis**

In the first stage, the standard method for two-time-scale systems is applied in which the previously derived Lyapunov functions for the slow and fast subsystems, that is, V_S and V_F, respectively, must satisfy certain inequalities.

- *Isolated equilibrium of the origin for the* $\sum_{SF}$*-subsystem*

The origin $(\tilde{X}_0 = 0,\ \tilde{Y}_0 = 0)$ is an isolated equilibrium of the $\sum_{SF}$-subsystem

$$0 = \hat{g}_0(0,0) \tag{68}$$

$$0 = \tilde{f}_0\left(0,0,\tilde{h}_0(\tilde{X}_0)\right) \tag{69}$$

moreover, $\tilde{Y}_0 = \tilde{g}_0(\tilde{X}_0)$ is the root of $0 = \hat{g}_0(\tilde{X}_0, \tilde{Y}_0)$ which vanishes at $\tilde{X}_0 = 0$, and $\left\|\tilde{g}_0(\tilde{X}_0)\right\| < \vartheta_1(\left\|\tilde{X}_0\right\|)$ where $\vartheta_1()$ is a κ function.

- *Reduced system condition for the* $\sum_{SF}$*-subsystem*

Using the Lyapunov function (67) and substituting $\rho\tilde{f}_0\left(\tilde{X}_0, \tilde{Y}_0, \tilde{h}_0(\tilde{X}_0)\right)$ yields

$$\left(\frac{\partial V_S(\tilde{X}_0)}{\partial \tilde{X}_0}\right)^T \rho\tilde{f}_0\left(\tilde{X}_0, \tilde{Y}_0, \tilde{h}_0(\tilde{X}_0)\right) \le -\alpha_1\phi_1^2(\tilde{X}_0) \tag{70}$$

where $\alpha_1 \le 1$ and $\phi_1(\tilde{X}_0) = \sqrt{\frac{q_{S1}}{2}\tilde{x}_{1,0}^2 + \frac{q_{S2}}{2}\tilde{x}_{2,0}^2 + \frac{q_{S3}}{2}\tilde{x}_{3,0}^2 + \frac{q_{S4}}{2}\tilde{x}_{4,0}^2}$.

- *Boundary-layer system condition for the* $\sum_{SF}$*-subsystem*

Since expression (62)–(65), it is seem that

$$\left(\frac{\partial V_{UF}(\tilde{X}_0, \tilde{Z}_0)}{\partial \tilde{Z}_0}\right)^T \rho\hat{H}(\tilde{X}_0, \tilde{Z}_0) \le -Q_{F2}\hat{y}_{2,0}^2 - Q_{F1}\hat{y}_{1,0}^2 \tag{71}$$

where Q_{F1} and Q_{F2} are defined as follows

$$Q_{F1} = \frac{q_{F1}}{2} - \frac{r_i q_{F2}}{2\psi'}\left(\left|\frac{4\left(r_i C_{pv}\left(\frac{\tilde{x}}{2,0} + \frac{i_{pv,0}}{C_{pv}}\right) - \frac{(\tilde{x}_{2,0}+x_{2,0}^*)}{2}\right)}{\left(2\psi' + \frac{C_{dc}k_2}{k_4 C_{pv}}\right)}\hat{y}_{1,0}\right| + \left|\hat{y}_{2,0}\right|\right) \tag{72}$$

$$Q_{F2} = \frac{q_{F2}}{2} - \frac{r_g q_{F2}}{4\psi'}\left(\left|\frac{4\left(r_i C_{pv}\left(\frac{\tilde{x}}{2,0} + \frac{i_{pv,0}}{C_{pv}}\right) - \frac{(\tilde{x}_{2,0}+x_{2,0}^*)}{2}\right)}{\left(2\psi' + \frac{C_{dc}k_2}{k_4 C_{pv}}\right)}\hat{y}_{1,0}\right| + \left|\hat{y}_{2,0}\right|\right) \tag{73}$$

For physical point of view and domain of working principle, it is supposed that all physical state variables are bounded in domain of interest, where $\tilde{X}_{0\min} \le \tilde{X}_0 \le \tilde{X}_{0\mathrm{Max}}$, $\tilde{Y}_{0\min} \le \tilde{Y}_0 \le \tilde{Y}_{0\mathrm{Max}}$, and $\tilde{Z}_{0\min} \le \tilde{Z}_0 \le \tilde{Z}_{0\mathrm{Max}}$. Therefore, Q_{F1} and Q_{F2} can take positives minimum possible values $Q_{F1,\min} > 0$ and $Q_{F2,\min} > 0$. This can be done by ensuring the appropriate selection of q_{F1}, q_{F2}, k_2, and k_4 as follows

$$\begin{cases} \dfrac{k_2}{k_4} > \dfrac{C_{pv}}{C_{dc}} \left(\dfrac{4\left| r_i C_{pv}\left(\dfrac{\tilde{x}}{2,0Max} + \dfrac{i_{pv,0}}{C_{pv}}\right) - \dfrac{(\tilde{x}_{2,0Max} + x^*_{2,0})}{2} \right|}{\left(\dfrac{2\psi'_{\min}}{r_g} - \hat{y}_{2,0\mathrm{Max}}\right)} \hat{y}_{1,0\mathrm{Max}} - 2\psi'_{\min} \right) \\ \dfrac{q_{F1}}{q_{F2}} > \dfrac{r_i}{\psi'_{\min}} \left(\dfrac{4\left|\left(r_i C_{pv}\left(\dfrac{\tilde{x}}{2,0Max} + \dfrac{i_{pv,0}}{C_{pv}}\right) - \dfrac{(\tilde{x}_{2,0Max} + x^*_{2,0})}{2} \right)\right|}{\left(2\psi'_{\min} + \dfrac{C_{dc}k_2}{k_4 C_{pv}}\right)} \hat{y}_{1,0\mathrm{Max}} + \hat{y}_{2,0\mathrm{Max}} \right) \end{cases} \tag{74}$$

Therefore, the boundary-layer system condition for the $\sum_{SF}$-subsystem is defined as follows

$$\left(\frac{\partial V_F}{\partial \tilde{Y}_0}\right)^T \rho \tilde{g}_0(\tilde{X}_0, \tilde{Y}_0) \le -\left(\sqrt{Q_{F2\min}\hat{y}^2_{2,0} + Q_{F1\min}\hat{y}^2_{1,0}}\right)^2 \le -\alpha_2 \phi_2^2(\hat{Y}_0) \tag{75}$$

where $\alpha_2 \le 1$ and $\phi_2(\hat{Y}_0) = \sqrt{Q_{F2\min}\hat{y}^2_{2,0} + Q_{F1\min}\hat{y}^2_{1,0}}$.

Now, we consider the composite Lyapunov function candidate of the $\sum_{SF}$-subsystem given as follow

$$V_{\mathrm{SF}} = (1-\eta_1)V_{\mathrm{S}}(\tilde{X}_0) + \eta_1 V_{\mathrm{F}}(\hat{Y}_0) \tag{76}$$

with $0 < \eta_1 < 1$. The derivative of V_{SF} presents new terms which represents the effect of the interconnection between the slow and fast dynamics. These interconnections are assumed to satisfy the following conditions

$$\left(\frac{\partial V_s(\tilde{X}_0)}{\partial \tilde{X}_0}\right)^T \rho\left(\tilde{f}_0\left(\tilde{X}_0, \tilde{Y}_0, \tilde{h}_0(\tilde{X}_0)\right) - \tilde{f}_0\left(\tilde{X}_0, \tilde{g}_0(\tilde{X}_0), \tilde{h}_0(\tilde{X}_0)\right)\right) \le \theta_1 \phi_2(\hat{Y}_0)\phi_1(\tilde{X}_0) \tag{77}$$

$$\left(\frac{\partial V_F(\hat{Y}_0)}{\partial \tilde{X}_0}\right)^T \rho \tilde{f}_0\left(\tilde{X}_0, \tilde{g}_0(\tilde{X}_0), \tilde{h}_0(\tilde{X}_0)\right) \le \theta_2 \phi_2(\hat{Y}_0)\phi_1(\tilde{X}_0) \tag{78}$$

$$\left(\frac{\partial V_F(\hat{Y}_0)}{\partial \tilde{X}_0}\right)^T \left[\rho \tilde{f}_0\left(\tilde{X}_0, \tilde{Y}_0, \tilde{h}_0(\tilde{X}_0)\right) - \rho \tilde{f}_0\left(\tilde{X}_0, \tilde{g}_0(\tilde{X}_0), \tilde{h}_0(\tilde{X}_0)\right)\right] \le \gamma_1 \phi_2^2(\hat{Y}_0) \tag{79}$$

where the constants θ_1, θ_2, and γ_1 are non-negative.

Therefore, from the singular perturbation theory (e.g. Theorem 11.3 in Ref. [19]), it follows that the derivative of V_{SF} is negative-definite for

$$\varepsilon_2 < \varepsilon_2^{\otimes} \quad \textit{for} \quad (\eta_1 < \eta_1^{\otimes}) \tag{80}$$

$$\text{with } \varepsilon_2^{\otimes} = \frac{\alpha_1\alpha_2}{\alpha_1\gamma_1 + \theta_1\theta_2} \text{ and } \eta_1^{\otimes} = \frac{\theta_1}{\theta_1 + \theta_2}. \tag{81}$$

$\sum SFU$ Stability Analysis:

The stability of full system is analyzed now. In this step, the results obtained in above section will be used in order to prove the asymptotic stability properties of the full $\sum_{SFU}$-system, which, for convenience, is first rewritten as

$$\dot{\tilde{\chi}}_0 = \tilde{F}_0(\tilde{\chi}_0, \tilde{Z}_0) \tag{82}$$

$$\varepsilon_1\varepsilon_2\dot{\tilde{Z}}_0 = \hat{h}_0(\tilde{\chi}_0, \tilde{Z}_0) \tag{83}$$

where $\tilde{\chi}_0 = \begin{pmatrix} \tilde{x}_{1,0} & \tilde{x}_{2,0} & \tilde{x}_{3,0} & \tilde{x}_{4,0} & \tilde{y}_{1,0} & \tilde{y}_{2,0} \end{pmatrix}^T$, similarly to the $\sum_{SF}$ general asymptotic stability analysis presented in the above section, the $\sum_{SFU}$-system is treated like a two-time-scale singularly perturbed system, where the $\sum_{SF}$-subsystem is treated as the new slow augmented reduced order. The stability of the full system implicates that the previously derived Lyapunov functions for the new slow and new fast subsystems, that is, $V_{SF}(\tilde{\chi}_0)$ and $V_U(\hat{Z}_0)$, respectively, must satisfy certain inequalities.

- *Isolated equilibrium of the origin for the $\sum_{SFU}$-subsystem*

The origin $(\tilde{\chi}_0 = 0, \tilde{Z}_0 = 0)$ is an isolated equilibrium of the $\sum_{SFU}$-subsystem

$$0 = \tilde{F}_0(0, 0) \tag{84}$$

$$0 = \hat{h}_0(0, 0) \tag{85}$$

Moreover, $\tilde{Z}_0 = \tilde{h}_0(\tilde{\chi}_0)$ is the unique root of $0 = \hat{h}_0(\tilde{\chi}_0, \tilde{Z}_0)$, which vanishes at $\tilde{\chi}_0 = 0$, and $\left\|\tilde{h}_0(\tilde{\chi}_0)\right\| < \vartheta_2(\|\tilde{\chi}_0\|)$ where $\vartheta_2()$ is a κ function.

- *Reduced system condition for the $\sum_{SFU}$-subsystem*

Using Lyapunov function candidate $V_{SF}(\tilde{\chi}_0)$ given by expression (76), it is easily shown that

$$\left(\frac{\partial V_{SF}(\tilde{\chi}_0)}{\partial \tilde{\chi}_0}\right)^T = \left(\frac{\partial V_{SF}(\tilde{\chi}_0)}{\partial \tilde{x}_{1,0}} \quad \frac{\partial V_{SF}(\tilde{\chi}_0)}{\partial \tilde{x}_{2,0}} \quad \frac{\partial V_{SF}(\tilde{\chi}_0)}{\partial \tilde{x}_{3,0}} \quad \frac{\partial V_{SF}(\tilde{\chi}_0)}{\partial \tilde{x}_{4,0}} \quad \frac{\partial V_{SF}(\tilde{\chi}_0)}{\partial \tilde{y}_{1,0}} \quad \frac{\partial V_{SF}(\tilde{\chi}_0)}{\partial \tilde{y}_{2,0}}\right)^T \tag{86}$$

Recalling that $\hat{Y}_0 = \tilde{Y}_0 - \tilde{g}_0(\tilde{X}_0)$, and since (61)–(65), one gets

$$\frac{\partial p_{F1}}{\partial \tilde{x}_{2,0}} = \left[\frac{r_g C_{pv}\left(2r_i C_{pv}\left(\frac{\tilde{x}}{T_2}+\frac{i_{pv}}{C_{pv}}\right)-\left(\tilde{x}_{2,0}+x_{2,0}^*\right)\right)^2 \left(\frac{2C_{pv}}{T_2}\left(1-\frac{r_i C_{pv}}{T_2}\right)\tilde{x}_2 + C_{pv}\left(\frac{x_{2,0}^*}{T_2}+\frac{i_{pv}}{C_{pv}}-\frac{2r_i i_{pv}}{T_2}\right)\right)}{k_2 \psi'^3\left(\psi' + \left(C_{dc}k_2/2k_4C_{pv}\right)\right)^2} \right.$$

$$\left. \times\left(2\psi' + \left(C_{dc}k_2/2k_4C_{pv}\right)\right) - \frac{2C_{pv}\left(2r_i C_{pv}\left(\frac{\tilde{x}}{T_2}+\frac{i_{pv}}{C_{pv}}\right)-\left(\tilde{x}_{2,0}+x_{2,0}^*\right)\right)\left(\frac{2r_iC_{pv}}{T_2}-1\right)}{k_2\psi'\left(\psi' + \left(C_{dc}k_2/2k_4C_{pv}\right)\right)} \right] q_{F2} = A_1 q_{F2}$$

$$\frac{\partial p_{F1}}{\partial \tilde{x}_{4,0}} = \frac{r_g C_{dc} C_{pv}\left(2r_i C_{pv}\left(\frac{\tilde{x}}{T_2}+\frac{i_{pv}}{C_{pv}}\right)-\left(\tilde{x}_{2,0}+x_{2,0}^*\right)\right)^2\left(2\psi' + \left(C_{dc}k_2/2k_4C_{pv}\right)\right)}{2k_2T_4\psi'^3\left(\psi' + \left(C_{dc}k_2/2k_4C_{pv}\right)\right)^2} q_{F2} = A_2 q_{F2}$$

$$\frac{\partial p_{F2}}{\partial \tilde{x}_{2,0}} = \frac{C_{dc}C_{pv}r_g\left(\frac{2}{T_2}\left(1-\frac{r_iC_{pv}}{T_2}\right)\tilde{x}_{2,0} + \left(\frac{x_{2,0}^*}{T_2}+\frac{i_{pv}}{C_{pv}}-\frac{2r_ii_{pv}}{T_2}\right)\right)}{4k_4\psi'^3} q_{F2} = A_3 q_{F2},$$

$$\frac{\partial p_{F2}}{\partial \tilde{x}_{4,0}} = \frac{r_g C_{dc}^2}{4T_4k_4\psi'^3} q_{F2} = A_4 q_{F2},$$

$$\frac{\partial p_{F3}}{\partial \tilde{x}_{4,0}} = -\frac{C_{dc}^2 r_g\left(2r_i C_{pv}\left(\frac{\tilde{x}}{T_2}+\frac{i_{pv}}{C_{pv}}\right)-\left(\tilde{x}_{2,0}+x_{2,0}^*\right)\right)\left(2\psi' + \left(C_{dc}k_2/2k_4C_{pv}\right)\right)}{4k_4T_4\psi'^3\left(\psi' + \left(C_{dc}k_2/2k_4C_{pv}\right)\right)^2} q_{F2} = A_6 q_{F2},$$

$$\frac{\partial p_{F3}}{\partial \tilde{x}_{2,0}} = \left[-\frac{C_{dc}r_g\left(2r_i C_{pv}\left(\frac{\tilde{x}}{T_2}+\frac{i_{pv}}{C_{pv}}\right)-\left(\tilde{x}_{2,0}+x_{2,0}^*\right)\right)\left(2\frac{C_{pv}}{T_2}\left(1-\frac{r_iC_{pv}}{T_2}\right)\tilde{x}_{2,0} + C_{pv}\left(\frac{x_{2,0}^*}{T_2}+\frac{i_{pv}}{C_{pv}}-\frac{2r_ii_{pv}}{T_2}\right)\right)}{4k_4\psi'^3\left(\psi' + \left(C_{dc}k_2/2k_4C_{pv}\right)\right)} \right.$$

$$\left. \times\left(2\psi' + \left(C_{dc}k_2/2k_4C_{pv}\right)\right) + \frac{\left(\frac{2r_iC_{pv}}{T_2}-1\right)C_{dc}}{4k_4\psi'\left(\psi' + \left(C_{dc}k_2/2k_4C_{pv}\right)\right)} \right] q_{F2} = A_5 q_{F2}$$

$$p_{F1}\frac{\partial \hat{y}_{1,0}}{\partial \tilde{x}_{2,0}} = \frac{C_{pv}^2}{2k_2T_2} q_{F1} + \frac{C_{pv}^2\left(2r_i C_{pv}\left(\frac{\tilde{x}}{T_2}+\frac{i_{pv}}{C_{pv}}\right)-\left(\tilde{x}_{2,0}+x_{2,0}^*\right)\right)^2}{2k_2T_2\psi'\left(\psi' + \left(C_{dc}k_2/2k_4C_{pv}\right)\right)} q_{F2} = A_{13}q_{F1} + A_7 q_{F2}$$

$$p_{F2}\frac{\partial \hat{y}_{2,0}}{\partial \tilde{x}_{2,0}} = \frac{C_{dc}\left(2\frac{C}{T_2}\left(1-\frac{r_iC_{pv}}{T_2}\right)\tilde{x}_{2,0} + C_{pv}\left(\frac{x_{2,0}^*}{T_2}+\frac{i_{pv}}{C_{pv}}-\frac{2r_ii_{pv}}{T_2}\right)\right)}{4k_4\psi'^2} q_{F2} = A_8 q_{F2}$$

$$p_{F3}\frac{\partial \hat{y}_{1,0}}{\partial \tilde{x}_{2,0}} = -\frac{C_{pv}C_{dc}\left(2r_i C_{pv}\left(\frac{\tilde{x}}{T_2}+\frac{i_{pv}}{C_{pv}}\right)-\left(\tilde{x}_{2,0}+x_{2,0}^*\right)\right)}{4T_2k_4\psi'\left(\psi' + \left(C_{dc}k_2/2k_4C_{pv}\right)\right)} q_{F2} = A_9 q_{F2},\ p_{F3}\frac{\partial \hat{y}_{1,0}}{\partial \tilde{x}_{4,0}} = 0$$

$$p_{F3}\frac{\partial \hat{y}_{2,0}}{\partial \tilde{x}_{2,0}} = -\frac{C_{dc}\left(2r_i C_{pv}\left(\frac{\tilde{x}}{T_2}+\frac{i_{pv}}{C_{pv}}\right)-\left(\tilde{x}_{2,0}+x_{2,0}^*\right)\right)\left(2\frac{C_{pv}}{T_2}\left(1-\frac{r_iC_{pv}}{T_2}\right)\tilde{x}_{2,0} + C_{pv}\left(\frac{x_{2,0}^*}{T_2}+\frac{i_{pv}}{C_{pv}}-\frac{2r_ii_{pv}}{T_2}\right)\right)}{4k_4\psi'^2\left(\psi' + \left(C_{dc}k_2/2k_4C_{pv}\right)\right)} q_{F2} = A_{10} q_{F2}$$

$$p_{F1}\frac{\partial \hat{y}_{1,0}}{\partial \tilde{x}_{4,0}} = 0,\ p_{F3}\frac{\partial \hat{y}_{2,0}}{\partial \tilde{x}_{4,0}} = -\frac{C_{dc}^2\left(2r_i C_{pv}\left(\frac{\tilde{x}}{T_2}+\frac{i_{pv}}{C_{pv}}\right)-\left(\tilde{x}_{2,0}+x_{2,0}^*\right)\right)}{4T_4k_4\psi'^2\left(\psi' + \left(C_{dc}k_2/2k_4C_{pv}\right)\right)} q_{F2} = A_{12} q_{F2},$$

$p_{F2}\frac{\partial \hat{y}_{2,0}}{\partial \tilde{x}_{4,0}} = \frac{C_{dc}^2}{4k_4T_4\psi'^2}q_{F2} = A_{11}q_{F2}$.

Substituting the expressions of $\rho\tilde{F}_0\left(\tilde{\chi}_0, \tilde{h}_0(\tilde{\chi}_0)\right)$ and by using absolute value version of Young's inequality $|ab| \leq \left|(a^2+b^2)/2\right|$, it shows that

$$\left(\frac{\partial V_1(\tilde{\chi}_0)}{\partial \tilde{\chi}_0}\right)^T \rho F_0\left(\tilde{\chi}_0, \tilde{h}_0(\tilde{\chi}_0)\right) \leq -\mathrm{M}_1\tilde{x}_{1,0}^2 - \mathrm{M}_2\tilde{x}_{2,0}^2 - \mathrm{M}_3\tilde{x}_{3,0}^2 \\ -\mathrm{M}_4\tilde{x}_{4,0}^2 - \mathrm{M}_5\hat{y}_{1,0}^2 - \mathrm{M}_6\hat{y}_{2,0}^2 \tag{87}$$

where $\mathrm{M}_1 = (1-\eta_1)\frac{q_{S1}}{2}$, $\mathrm{M}_2 = (1-\eta_1)\frac{q_{S2}}{2} - \frac{(1-\eta_1)}{2}\left(\frac{T_2}{2}q_{S2}\right)^2 - \frac{\eta_1}{2}\left(-\frac{1}{T_2}\right)^2$, $\mathrm{M}_3 = -(1-\eta_1)\frac{q_{S3}}{2}$, $\mathrm{M}_4 = (1-\eta_1)\frac{q_{S4}}{2} - \frac{(1-\eta_1)}{2}\left(\frac{T_4}{2}q_{S4}\right)^2 - \frac{\eta_1}{2}\left(-\frac{1}{T_4}\right)^2$, $M_5 = \frac{q_{F1}}{2} - M_5' - \frac{(1-\eta_1)}{2C_{pv}^2}$, and $M_6 = \frac{q_{F2}}{2} - M_6' - \frac{\eta_1}{2}\left(-\frac{2\psi'}{C_{dc}} - \frac{r_g}{C_{dc}}\hat{y}_{2,0}\right)^2_{Max}$

$$M_5' = \eta_1\left|\frac{A_{13}q_{F1} + \left(\frac{A}{2}\hat{y}_{1,0} + A_5\hat{y}_{2,0} + (A_7+A_{10})\right)q_{F2}}{C_{pv}}\right|_{Max} + \eta_1 q_{F2}\left|\frac{\frac{A}{2}\hat{y}_{2,0} + A_8 + A_9}{2C_{pv}}\right|_{Max}$$

$$+\eta_1\frac{q_{F2}}{2}\left(\frac{2(\tilde{x}_{2,0}+x_{2,0}^*)}{C_{dc}} + \frac{2r_i}{C_{dc}}\hat{y}_{1,0}\right)^2_{Max} + \frac{\eta_1}{2}\left(\left(\frac{A_1}{2}\hat{y}_{1,0} + A_5\hat{y}_{2,0} + (A_7+A_{10})\right)q_{F2} + A_{13}q_{F1}\right)^2_{Max} + \eta_1 q_{F2}\left|\left(\frac{A_2}{2}\hat{y}_{1,0} + A_6\hat{y}_{2,0} + A_{12}\right)\left(\frac{2(\tilde{x}_{2,0}+x_{2,0}^*)}{C_{dc}} + \frac{2r_i}{C_{dc}}\hat{y}_{1,0}\right)\right|_{Max}$$

$$+\eta_1\frac{q_{F2}}{2}\left|\left(\frac{A_2}{2}\hat{y}_{1,0} + A_6\hat{y}_{2,0} + A_{12}\right)\left(-\frac{2\psi'}{C_{dc}} - \frac{r_g}{C_{dc}}\hat{y}_{2,0}\right)\right|_{Max} + \eta_1 q_{F2}$$

$$\times\left|\left(\frac{A_4}{2}\hat{y}_{2,0} + A_{11}\right)\left(\frac{2(\tilde{x}_{2,0}+x_{2,0}^*)}{C_{dc}} + \frac{2r_i}{C_{dc}}\hat{y}_{1,0}\right)\right|_{Max} + \eta_1\left(\left(\frac{A_2}{2}\hat{y}_{1,0} + A_6\hat{y}_{2,0} + A_{12}\right)q_{F2}\right)^2_{Max}$$

$$+\eta_1 q_{F2}\left|-\frac{r_i}{2\psi'}\hat{y}_{2,0} + \frac{\left(2r_iC_{pv}\left(\frac{\tilde{x}}{T_2} + \frac{i_{pv,0}}{C_{pv}}\right) - (\tilde{x}_{2,0}+x_{2,0}^*)\right)r_i}{2\psi'(\psi' + (C_{dc}k_2/2k_4C_{pv}))}\hat{y}_{1,0}\right|_{Max}$$

$$M_6' = \eta_1\frac{q_{F2}}{2}\left|\frac{\frac{A}{2}\hat{y}_{2,0} + A_8 + A_9}{2C_{pv}}\right|_{Max} + \frac{\eta_1}{2}\left(\left(\frac{A_3}{2}\hat{y}_{2,0} + A_8 + A_9\right)q_{F2}\right)^2_{Max}$$

$$+\frac{\eta_1}{2}\left(\left(+\frac{A_4}{2}\hat{y}_{2,0} + A_{11}\right)q_{F2}\right)^2_{Max} + \eta_1 q_{F2}\left|\left(\frac{A_4}{2}\hat{y}_{2,0} + A_{11}\right)\left(-\frac{2\psi'}{C_{dc}} - \frac{r_g}{C_{dc}}\hat{y}_{2,0}\right)\right|_{Max}$$

$$+\eta_1\frac{q_{F2}}{2}\left|\left(\frac{A_2}{2}\hat{y}_{1,0} + A_6\hat{y}_{2,0} + A_{12}\right)\left(-\frac{2\psi'}{C_{dc}} - \frac{r_g}{C_{dc}}\hat{y}_{2,0}\right)\right|_{Max}$$

$$+\eta_1\frac{q_{F2}}{2}\left|\left(+\frac{A_4}{2}\hat{y}_{2,0} + A_{11}\right)\left(\frac{2(\tilde{x}_{2,0}+x_{2,0}^*)}{C_{dc}} + \frac{2r_i}{C_{dc}}\hat{y}_{1,0}\right)\right|_{Max}$$

$$+\eta_1 q_{F2}\left|-\frac{r_g}{4\psi'}\hat{y}_{2,0} + \frac{r_g\left(2r_iC_{pv}\left(\frac{\tilde{x}}{T_2} + \frac{i_{pv,0}}{C_{pv}}\right) - (\tilde{x}_{2,0}+x_{2,0}^*)\right)}{4\psi'(\psi' + (C_{dc}k_2/2k_4C_{pv}))}\hat{y}_{1,0}\right|_{Max}$$

Now, it needs to ensure that $-M_1\tilde{x}_{1,0}^2 - M_2\tilde{x}_{2,0}^2 - M_3\tilde{x}_{3,0}^2 - M_4\tilde{x}_{4,0}^2 - M_5\hat{y}_{1,0}^2 - M_6\hat{y}_{2,0}^2 < 0$. This can be done by ensuring the appropriate selection of q_{S1}, q_{S2}, q_{S3}, q_{S4}, q_{F1}, and q_{F2}, such that q_{S1} and q_{S3} have sufficient large positive values, q_{F1} and q_{F2} have sufficient small positive values, and q_{S2} and q_{S4} are limited as follows $\frac{1-\sqrt{1-\frac{\eta}{(1-\eta_1)}}}{\frac{T_2^2}{2}} < q_{S2} < \frac{1+\sqrt{1-\frac{\eta}{(1-\eta_1)}}}{\frac{T_2^2}{2}}$, $\frac{1-\sqrt{1-\frac{\eta}{(1-\eta_1)}}}{\frac{T_4^2}{2}} < q_{S4} < \frac{1+\sqrt{1-\frac{\eta}{(1-\eta_1)}}}{\frac{T_4^2}{2}}$ (for $\eta_1 < \frac{1}{3}$). Therefore, the reduced system condition for the $\sum_{SFU}$ system is defined as follows

$$\left(\frac{\partial V_1(\tilde{\chi}_0)}{\partial \tilde{\chi}_0}\right)^T \rho F_0\left(\tilde{\chi}_0, \tilde{h}_0(\tilde{\chi}_0)\right) \le -\alpha_3 \phi_3^2(\tilde{\chi}_0) \tag{88}$$

where $\alpha_3 \le 1$ and $\phi_3(\tilde{\chi}_0) = \sqrt{M_1\tilde{x}_{1,0}^2 + M_2\tilde{x}_{2,0}^2 + M_3\tilde{x}_{3,0}^2 + M_4\tilde{x}_{4,0}^2 + M_5\hat{y}_{1,0}^2 + M_6\hat{y}_{2,0}^2}$.

- Boundary-layer system condition for the $\sum_{SFU}$-system

Using the Lyapunov function given by expression (58)–(60), and substituting $\rho\hat{h}_0(\tilde{X}_0, \tilde{Z}_0)$ of expression (57), one obtains

$$\left(\frac{V_u(\tilde{\chi}_0, \tilde{Z}_0)}{\partial \tilde{Z}_0}\right)\rho\hat{h}_0(\tilde{\chi}_0, \tilde{Z}_0) = \frac{q_{U1}}{2}\hat{z}_{1,0}^2 + \frac{q_{U2}}{2}\hat{z}_{2,0}^2 \le -\alpha_4\phi_4^2(\hat{Z}_0) \tag{89}$$

where $\alpha_4 \le 1$ and $\phi_4(\hat{Z}_0) = \sqrt{\frac{q_{U1}}{2}\hat{z}_{1,0}^2 + \frac{q_{U2}}{2}\hat{z}_{2,0}^2}$.

Now, we consider the composite Lyapunov function candidate of the full system given as follow

$$V_{SFU}(\tilde{\chi}_0, \hat{z}_0) = (1-\eta_2)V_{SF}(\tilde{\chi}_0) + \eta_2 V_U(\hat{z}_0) \tag{90}$$

with $0 < \eta_2 < 1$. The derivative of $V_{SFU}(\tilde{\chi}_0, \hat{z}_0)$ presents new terms, which represents the effect of the interconnection between the slow and fast dynamics. These interconnections are assumed to satisfy the following conditions:

$$\left(\frac{\partial V_{SF}(\tilde{\chi}_0)}{\partial \tilde{\chi}_0}\right)^T \rho\left(\tilde{F}_0(\tilde{\chi}_0, \tilde{Z}_0) - \tilde{F}_0\left(\tilde{\chi}_0, \tilde{h}_0(\tilde{\chi}_0)\right)\right) \le \theta_3\phi_4(\hat{Z}_0)\phi_3(\tilde{\chi}_0) \tag{91}$$

$$\left(\frac{\partial V_u(\hat{Z}_0)}{\partial \tilde{\chi}_0}\right)^T \rho\tilde{F}_0\left(\tilde{\chi}_0, \tilde{h}_0(\tilde{\chi}_0)\right) \le \theta_4\phi_4(\hat{Z}_0)\phi_3(\tilde{\chi}_0) \tag{92}$$

$$\left(\frac{\partial V_u(\hat{Z}_0)}{\partial \tilde{\chi}_0}\right)^T \rho\left[\tilde{F}_0(\tilde{\chi}_0, \tilde{Z}_0) - \tilde{F}_0\left(\tilde{\chi}_0, \tilde{h}_0(\tilde{\chi}_0)\right)\right] \le \gamma_2\phi_4^2(\hat{Z}_0) \tag{93}$$

the constants θ_3, θ_4, and γ_2 are non-negative. Therefore, from the singular perturbation theory (e.g. Theorem 11.3 in Refs. [19, 23, 24]), it follows that the derivative of V_{SFU} is negative-definite when

$$\varepsilon_2 < \varepsilon_2^{\otimes} \quad for \ (\eta_1 < \eta_1^{\otimes}) \tag{94}$$

with $\varepsilon_2^{\otimes} = \frac{\alpha_1\alpha_2}{\alpha_1\gamma_1 + \theta_1\theta_2}$ and $\eta_1^{\otimes} = \frac{\theta_1}{\theta_1 + \theta_2}$.

Therefore, it can be inferred that the equilibrium $X_0^* = \left(x_{1,0}^* x_{2,0}^* x_{3,0}^* x_{4,0}^*\right)^T$, $Y_0^* = \left(y_{1,0}^* y_{2,0}^*\right)^T$, and $Z_0^* = \left(z_{1,0}^* z_{2,0}^*\right)^T$ of full system, is asymptotically stable for all $\varepsilon_i < \varepsilon_i^{\otimes}$ $(i = 1, 2)$. Therefore, Part 2 immediately follows from the averaging theory (e.g. Theorem 10.4 in Ref. [19]). Theorem is established.

Author details

Youssef Mchaouar[1]*, Abdelmajid Abouloifa[1], Ibtissam Lachkar[2] and Mohammed Fettach[1]

1 LTI Lab, Faculty of Sciences Ben M'sik, University Hasan II of Casablanca, Casablanca, Morocco

2 LISER Lab, ENSEM of Casablanca, University Hasan II of Casablanca, Casablanca, Morocco

*Address all correspondence to: uns1mchaouar@gmail.com

References

[1] Akorede MF, Hizam H, Pouresmaeil E. Distributed energy resources and benefits to the environment. Renewable and Sustainable Energy Reviews. 2010; **14**(2):724-734

[2] Kerekes T, Liserre M, Mastromauro RA, Dell'Aquila A. A single-phase voltage-controlled grid-connected photovoltaic system with power quality conditioner functionality. IEEE Transactions on Industrial Electronics. 2009;**56**(11):4436-4444

[3] Gu B, Dominic J, Lai J, Chen C, La Bella T, Chen B. High reliability and efficiency single-phase transformerless inverter for grid-connected photovoltaic systems. IEEE Transactions on Power Electronics. 2013;**28**(5):2235-2245

[4] Yang Y, Blaabjerg F, Zou Z. Benchmarking of grid fault modes in single-phase grid-connected photovoltaic systems. IEEE Transactions on Industry Applications. 2013;**49**(5): 2167-2176

[5] Reza Reisi A, Hassan Moradi M, Jamasb S. Classification and comparison of maximum power point tracking techniques for photovoltaic system: A review. Renewable and Sustainable Energy Reviews. 2013;**19**:433-443

[6] Lin C-H, Huang C-H, Du Y-C, Chen J-L. Maximum photovoltaic power tracking for the PV array using the fractional-order incremental conductance method. Applied Energy. 2011;**88**(12):4840-4847

[7] Sera D, Mathe L, Kerekes T, Spataru SV, Teodorescu R. On the perturb-and-observe and incremental conductance MPPT methods for PV systems. IEEE Journal of Photovoltaics. 2013;**3**(3):1070-1078

[8] Ishaque K, Salam Z, Lauss G. The performance of perturb and observe and incremental conductance maximum power point tracking method under dynamic weather conditions. Applied Energy. 2014;**119**:228-236

[9] Gounden NA, Ann Peter S, Nallandula H, Krithiga S. Fuzzy logic controller with MPPT using line-commutated inverter for three-phase grid-connected photovoltaic systems. Renewable Energy. 2009;**34**(3):909-915

[10] Sefa I, Altin N, Ozdemir S, Kaplan O. Fuzzy PI controlled inverter for grid interactive renewable energy systems. IET Renewable Power Generation. 2015;**9**(7):729-738

[11] Ciobotaru M, Teodorescu R, Blaabjerg F. Control of single-stage single-phase PV inverter. In: 2005 European Conference on Power Electronics and Applications. 2005. p. 10

[12] Aouadi C et al. Multi loop based control of photovoltaic system connected to the single phase grid. In: 2016 International Renewable and Sustainable Energy Conference (IRSEC). 2016. pp. 479-483

[13] Menadi A, Abdeddaim S, Ghamri A, Betka A. Implementation of fuzzy-sliding mode based control of a grid connected photovoltaic system. ISA Transactions. 2015;**58**:586-594

[14] Mahmud MA, Pota HR, Hossain MJ. Nonlinear current control scheme for a single-phase grid-connected photovoltaic system. IEEE Transactions on Sustainable Energy. 2014;**5**(1): 218-227

[15] Mchaouar Y et al. Nonlinear control of single-phase grid-connected photovoltaic systems via singular perturbation. In: 2017 IEEE International Conference on Smart

Energy Grid Engineering (SEGE). 2017. pp. 210-216

[16] Hassaine L, OLias E, Quintero J, Salas V. Overview of power inverter topologies and control structures for grid connected photovoltaic systems. Renewable and Sustainable Energy Reviews. 2014;**30**:796-807

[17] Ravi A, Manoharan PS, Vijay Anand J. Modeling and simulation of three phase multilevel inverter for grid connected photovoltaic systems. Solar Energy. 2011;**85**(11):2811-2818

[18] Luo A, Chen Y, Shuai Z, Tu C. An improved reactive current detection and power control method for single-phase photovoltaic grid-connected DG system. IEEE Transactions on Energy Conversion. 2013;**28**(4):823-831

[19] Khalil HK. Nonlinear Systems. 3rd ed. Upper Saddle River, N.J: Pearson; 2001

[20] Kokotovic P, Khalil H, O'Reilly J. Society for Industrial and Applied Mathematics; 1999

[21] Yurkevich VD. Design of nonlinear control systems with the highest derivative in feedback. Singapore: World Scientific; 2004. xx + 352 pp. ISBN: 981-238-899-0

[22] Mchaouar Y et al. Nonlinear control of single-phase shunt active power filters with theoretical analysis via singular perturbation. In: 2018 5th International Conference on Electrical and Electronic Engineering (ICEEE). 2018. pp. 67-72

[23] Esteban S, Vázquez R, Aracil J. Singular perturbation stability analysis for a three-time-scale autonomous helicopter; 2012

[24] Esteban S, Gordillo F, Aracil J. Three-time scale singular perturbation control and stability analysis for an autonomous helicopter on a platform. International Journal of Robust and Nonlinear Control. 2013;**23**(12): 1360-1392

[25] Abouloifa A, Giri F, Lachkar I, Chaoui FZ, Kissaoui M, Abouelmahjoub Y. Cascade nonlinear control of shunt active power filters with average performance analysis. Control Engineering Practice. 2014;**26**: 211-221

[26] Enrique JM, Durán E, Sidrach-de-Cardona M, Andújar JM. Theoretical assessment of the maximum power point tracking efficiency of photovoltaic facilities with different converter topologies. Solar Energy. 2007;**81**(1): 31-38

[27] Krein PT, Bentsman J, Bass RM, Lesieutre BL. On the use of averaging for the analysis of power electronic systems. IEEE Transactions on Power Electronics. 1990;**5**(2):182-190

Chapter 10

Battery Internal Fault Monitoring Based on Anomaly Detection Algorithm

Nassim Sabri, Abdelhalim Tlemçani and Aissa Chouder

Abstract

Battery internal faults are one of the major factors causing safety concern, performance degradation, and cost increases. To extend the lifetime of the battery and bring more security in the system, internal fault detection of solar battery is proposed in this paper using an unsupervised machine learning algorithm based on anomaly detection method. The advantages of adopting such a method consist of using unlabeled data that meet the battery case in the difficulty of obtaining the fault data. In contrast, healthy data can easily be obtained from the battery and therefore allows building the anomaly detection algorithm. The effectiveness of the proposed method is validated using a simulation platform of a stand-alone photovoltaic system developed in Matlab/Simulink that takes as system input a real profile of irradiance and temperature captured from the Centre de Development des Energies renewable (CDER), Algeria. The test results in real-time data show the ability of the proposed approach to detect the fault occurrence in the battery.

Keywords: anomaly detection, internal faults, battery, simulation, data

1. Introduction

The future development of stand-alone photovoltaic (SAPV) systems will know a progression in all countries, especially in remote areas and islands [1]. The battery is the main component in SAPV system, since it represents 43% of lifecycle costs for this system [2], where this lifecycle could significantly reduce in badly work conditions. To check continuously the status of battery and aging, monitoring of internal resistance is proposed in [3, 4] proposed an estimation of solar irradiation to monitor the overcharge and internal resistance in Battery. Other studies [5–8] focusing to assess the performance of the whole SAPV system using energy parameters. This latter provides an analysis of SAPV system, where the battery takes a part in this system. Fault diagnosis in SAPV system including the specific fault battery external short-circuit is proposed in [9, 10] using Artificial Neural Network, the study is experimentally validated an additionally integrated with the system using Matlab Graphical User Interface (GUI) [11]. Battery performance evaluation in PV-diesel hybrid system is proposed in [12] where the analysis requires many days and provides inaccurate results concerning the battery state.

The failure developed in battery, in particular, the internal faults could bring a difficulty to eliminate them by electrical protection elements such as fuse or circuit-breaker by reason of absence an appropriate selection of fuse for this circuit [13]. This will result in undetected faults that conduct the battery to work in low performance and decrease its lifecycle, furthermore, it increases the risk of fire hazard and even an explosion, without account the contamination of equipment's beside [13]. To solve this problem and overcome the limitation of existed battery fault analysis, this paper proposes internal faults detection of solar battery using anomaly detection technique. This method is commonly utilized for fault detection [14] specifically in the photovoltaic system [15, 16], where the proposed anomaly detection prediction model has the ability of development with only normal data and using the readily available battery voltage and current without requiring extra sensor circuit. Furthermore, it could recognize any non-specific failure in battery which makes them a promising choice for practical application.

The training and test for validation have been performed using data from the simulation platform of SAPV system, the battery is adequately represented in Matlab\SimPowerSystem to allows the creation of fault such as short-circuit and ground-fault. In addition, to make the simulation varies along with irradiance level and temperature, a real climatic measurements profile captured from CDER is used as input to estimate the output.

The test results of fault detection using anomaly detection method in real-time data show the capability of anomaly detection to recognize all the faults in the battery.

2. Modeling of SAPV system

A typical SAPV system as shown in **Figure 1** consists mainly of PV panels, Battery, load, and charge regulator [17]. The PV panels are the DC power source, and the battery is the storage element.

2.1 Output PV panel modeling

The output current and voltage of PV panel is obtained using the common one-diode model [18], in which the relationship I-V is given as follow:

$$I = I_{\text{ph}} - I_0 \left[\exp\left(\frac{V + R_s I}{n V_t} \right) - 1 \right] + \frac{V + R_s I}{R_{\text{sh}}} \tag{1}$$

where I_{ph} is the light generated current, I_0 is the reverse saturation current of diode, n is the diode ideality factor, R_s and R_{sh} are the series and shunt resistance of panel respectively, and V_t is the thermal voltage.

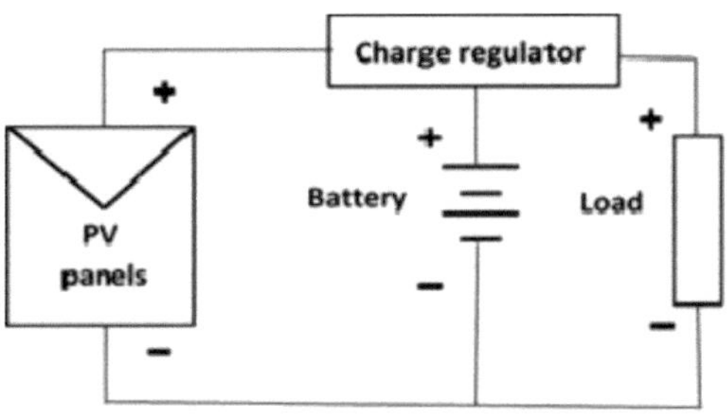

Figure 1.
Typical SAPV system.

2.2 Battery modeling

The charge and discharge models of lead-acid battery implemented in SimPowerSystems simulation environment [19] are summarized in the following:

2.2.1 Model for charge

$$V_{\text{batt}} = E_0 - R \cdot i - k \frac{Q}{it - 0.1 \cdot Q} \cdot i^* - k \frac{Q}{Q - it} \cdot it + \exp(t) \tag{2}$$

2.2.2 Model for discharge

$$V_{\text{batt}} = E_0 - R \cdot i - k \frac{Q}{Q - it} \cdot (it + i^*) + \exp(t) \tag{3}$$

where V_{batt} is the voltage of battery (V), E_0 is the constant voltage of battery (V), k is the polarization resistance (ohm), Q is the capacity of battery (Ah), it is the present battery charge (Ah), i is the current of battery (A), i^* is the filtered (A) current and exp. (t) is the exponential zone voltage (V).

2.3 Load and charge regulator representation

The load is represented by a resistance connected in parallel with the PV panels and battery. The charge regulator consists of a simple switch on/off placed in one side between PV panels and battery in our case.

3. Proposed fault detection method

Anomaly detection or known as novelty detection or outlier detection [20] is one the most machine learning technique used in fault detection [21], which aim to detect abnormalities or unusual operation that can come up, it makes the assumption that the data are distributed according to Gaussian (or Normal) distribution and this latter can be modeled based on two parameters: the mean μ and the variance σ^2.

Three phases are required to build an anomaly detection model, the first is training phase, where the Gaussian distribution is estimated by the parameters μ and σ^2, the second is validation phase, in this step, some threshold ε is selected as a limit of being an anomaly (outlier) compared to the Gaussian probability function, the third is testing phase, in which a test is performed to check the performance of the model. In the following, Mathematical equations behind this approach are given.

3.1 Gaussian distribution

Fit a model of Gaussian distribution from data relies on the assumption that huge number of data used is normal data. For each feature x_j (j = 1, … ,n) an estimation of Gaussian distribution parameters μ_j and σ_j^2. The Gaussian probability density is defined as:

$$P(x) = \frac{1}{\sqrt{2\pi}\sigma} \exp\left(-\frac{1}{2\sigma^2}(x - \mu)^2\right) \tag{4}$$

3.2 Estimation parameters for Gaussian distribution

The parameters of Gaussian distribution μ_j and σ_j^2 for the j-th feature are estimated respectively as follow:

$$\mu_j = \frac{1}{m}\sum_{i=1}^{m} x_j^{(i)} \tag{5}$$

$$\sigma_j^2 = \frac{1}{m}\sum_{i=1}^{m} \left(x_j^{(j)} - \mu_j\right)^2 \tag{6}$$

3.3 Choosing the value of threshold ε

To decide if new examples are anomalies or not, the procedure consists to compare the value of Gaussian probability density with a threshold or limit, if the probability is lower than a certain threshold then these examples are anomalies. However, the selection of the threshold ε could be done by the trial-error method using the F1 score (Eq. (7)) metric as criteria. As shown in **Figure 2**, the algorithm proceeds to try several values of ε, where the chosen value of ε corresponds to the maximum F1 score [14] defined below:

$$F1 = \frac{2 \cdot prec \cdot rec}{prec + rec} \tag{7}$$

where *prec* is the precision and *rec* is the recall, they can be obtained by:

$$prec = \frac{tp}{tp + fp} \tag{8}$$

$$rec = \frac{tp}{tp + fn} \tag{9}$$

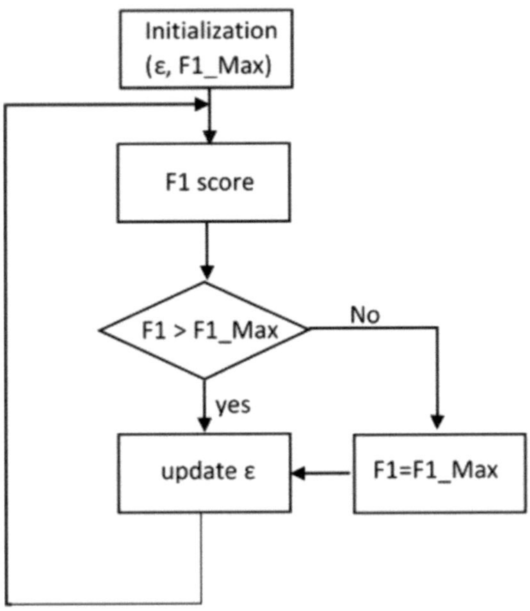

Figure 2.
Flowchart of the procedure to select the threshold.

Where *tp* is the true positive that means that the algorithm successfully classified as positive or anomaly, *fp* is the false positive that means that the algorithm incorrectly classify it as positive or anomaly, when it is not an anomaly, finally *fn* is the false negative that means that the algorithm wrongly classify it as not anomaly, while this sample is an anomaly.

4. Case study

A small SAPV system composed of 2 PV panels (212 Wp) connected with a battery of 1200 Ah and a load of 50 W is considered in this work. The PV panel used is Isofoton 106 W-12 V, where the parameters for this panel are obtained from [22]. Industrial lead-acid battery of 12 V contains six cells connected in series, in order to investigate internal faults in the battery, each cell is represented by a single battery of 2 V connected in series to model a real battery of 12 V, we point out that the cells are assumed has identical electrical characteristics.

The object of this paper is detecting internal faults that occur in the battery, where two faults are considered: ground fault and short-circuit (**Figure 3**), in which ground fault situate at three locations: upper, middle and lower, and short-circuit between cells consist of 1, 2, 3, 4 or 5 cells circuited. These faults are used to test and evaluate the detection approach.

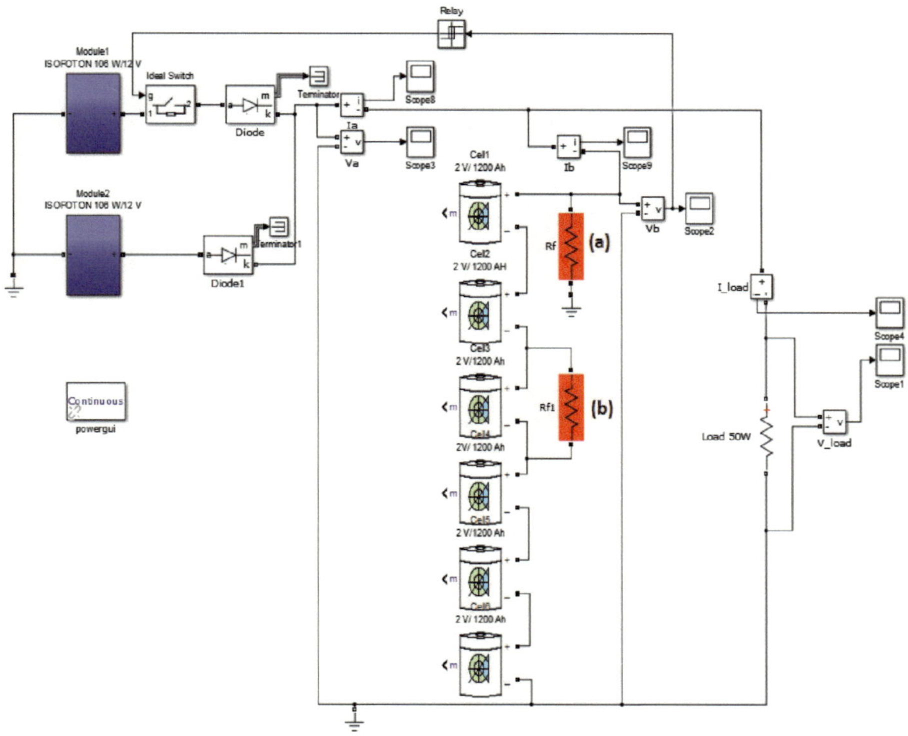

Figure 3.
SimPowerSystem model of SAPV system with battery.

5. Results and discussion

A real environmental measurement (irradiance and temperature) illustrated in **Figure 4**, taken from CDER, Bouzareah, Algeria, In which nine clear days are used as input to simulate the SAPV system under Matlab/Simulink.

In the following, some scenarios of faults are performed in order to test the capability of anomaly detection to recognize these faults using the last 4 days of data constituted of battery voltage and current, The Gaussian probability density is calculated and plotted with the threshold ε before and after occurring the fault.

In **Figure 5** a test for the Normal operation is realized, it can be seen that the probability does not drop below the threshold ε apart three false alarms noticed, which means there is no fault detected by the detector system.

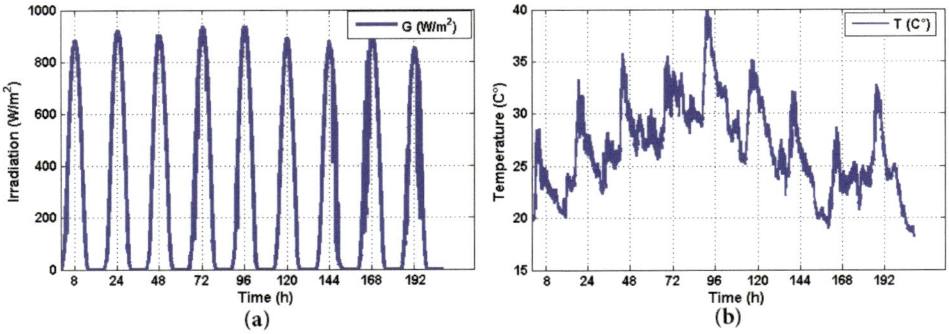

Figure 4.
Captured irradiance and temperature.

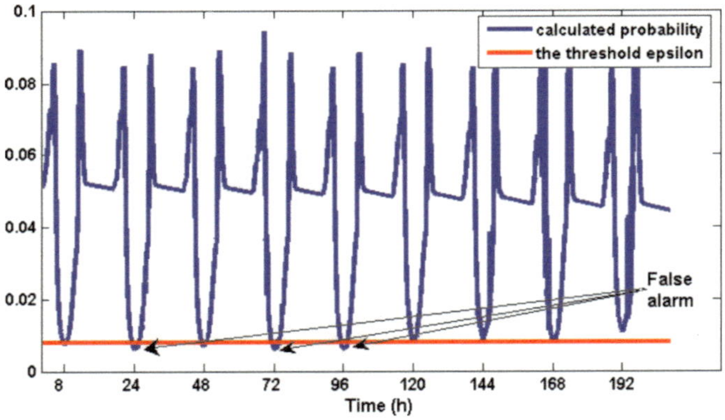

Figure 5.
Simulation test result: normal operation.

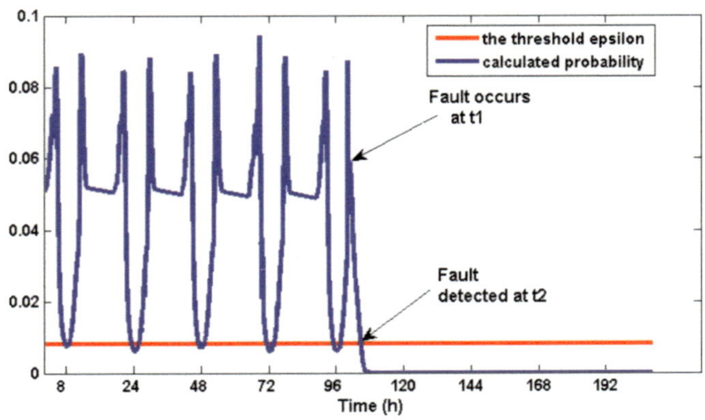

Figure 6.
Simulation test result: ground fault upper cell.

As illustrated in **Figure 6**, when a ground fault in the upper cell occurs, the probability decrease under the threshold ε after a certain time, and **Figure 7** shows a ground fault in the middle cell, which takes longer time than the previous fault before to be detected.

In **Figure 8** a short-circuit between 2 cells is created and as illustrated this fault take much time to descend under the threshold ε, while in **Figure 9** a similar fault is

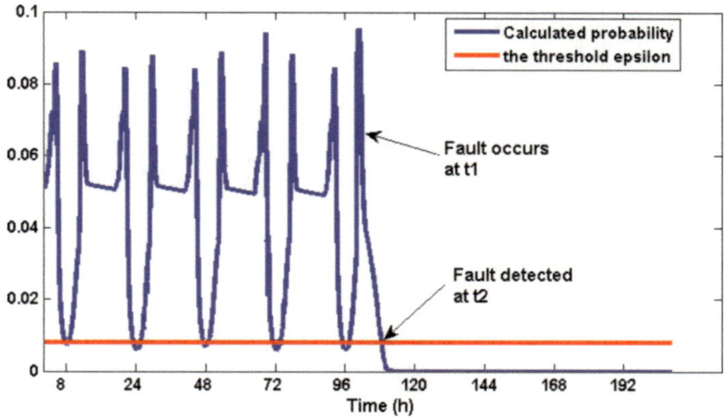

Figure 7.
Simulation test result: ground fault middle cell.

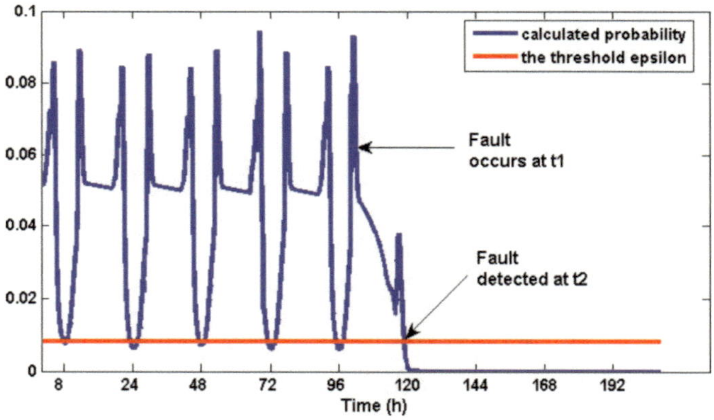

Figure 8.
Simulation test result: short-circuit of 2 cells.

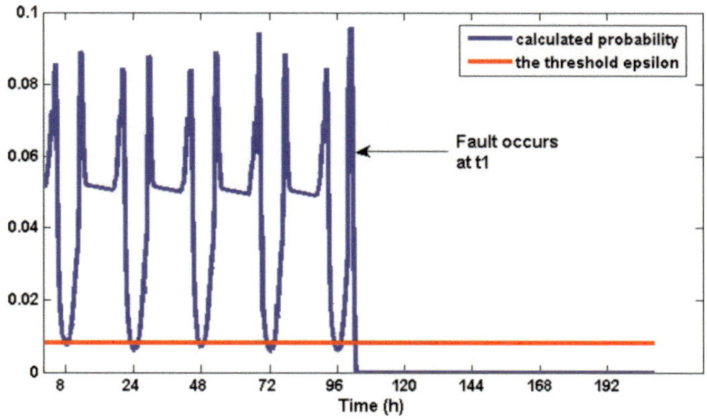

Figure 9.
Simulation test result: short-circuit of 2 cells ($Rf \approx 0$).

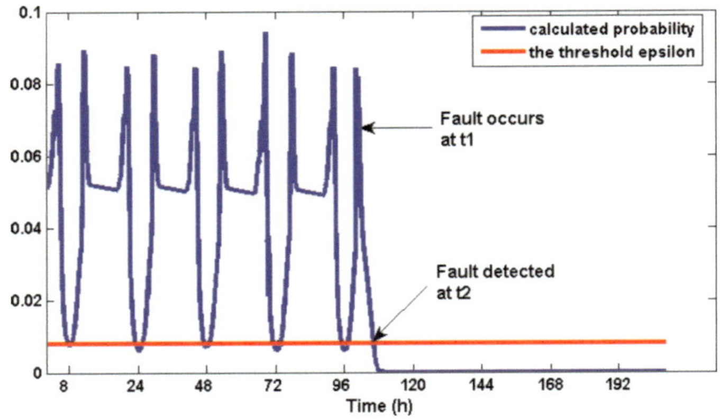

Figure 10.
Simulation test result: short-circuit of 5 cells.

created with Rf ≈ 0, then the probability drop below the threshold instantly and detected.

Figure 10 indicates the detection of short-circuit of 5 cells after a specific time of occurring this fault. From all these figures, it shows that all the faults are detected, whereas some faults take much time to detected than other, and this by the fact that the faulted voltage and current of Battery is reduced gradually and not immediately, and this depend on the external circuitry at moment of fault, in our case this is due to the value of Rf.

6. Conclusions

In this paper, a detection of battery cell interconnection fault is proposed using anomaly detection algorithm. The method has the benefits of using only the steady-state of battery and uses the easily available battery voltage and current to predict the internal fault in battery. Based on simulation data, the anomaly detection is developed and tested for validation, where in the simulation environment, the battery unit is viewed as series-connected battery cells. In this way, several typical faults such as internal short and ground fault are carried out. The proposed method is capable to effectively predict the battery internal faults, where the analysis finding reveal that as more battery cell are involved at faults or the fault circuit has negligible resistance, the fault detection becoming much faster to indicate the fault occurrence. The future work consists to implement this method on real battery as well as extend the application of battery fault detection to include battery EV and other appliances. Furthermore, another method based on a statistical approach to select the threshold ε in a better way would be proposed in the future, these techniques will perfectly manage to find the outlier of battery operation.

Author details

Nassim Sabri[1*], Abdelhalim Tlemçani[1] and Aissa Chouder[2]

1 Dr. Yahia Farés University, Laboratory of Electrical Engineering and Automatics (LREA), Médéa, Algeria

2 Mohamed Boudiaf University, Laboratory of Electrical Engineering (LGE), M'sila, Algeria

*Address all correspondence to: sabri_nassim@hotmail.com

References

[1] IEA-PVPS. Trends 2018 in Photovoltaics Applications. Report IEA PVPS T1-34:2018. 2018. pp. 73-91

[2] Luque A, Hegedus S. Handbook of Photovoltaic Science and Engineering. John Wiley Sons, Ltd; 1 Jan 2003;92-100

[3] Kim M, Hwang E. Monitoring the battery status for photovoltaic systems. Journal of Power Sources. 1997;**64**(97): 193-196

[4] Tadj M, Benmouiza K, Cheknane A. An innovative method based on satellite image analysis to check fault in a PV system lead-acid battery. Simulation Modelling Practice and Theory. 2014;**47** (14):236-247

[5] Mayer D, Heidenreich M. Performance analysis of stand alone PV systems from a rational use of energy point of view. In: 3rd World Conference on Photovolt Energy Conversion; 2003

[6] Muñoz FJ, Almonacid G, Nofuentes G, Almonacid F. A new method based on charge parameters to analyse the performance of stand-alone photovoltaic systems. Solar Energy Materials & Solar Cells. 2006;**90**(06): 1750-1763

[7] Muñoz FJ, Echbarthi I, Nofuentes G, Fuentes M, Aguilera J. Estimation of the potential array output charge in the performance analysis of stand-alone photovoltaic systems without MPPT (Case study: Mediterranean climate). Solar Energy. 2009;**83**(09):1985-1997

[8] Torres M, Muñoz FJ, Muñoz JV, Rus C. Online monitoring system for stand-alone photovoltaic applications—Analysis of system performance from monitored data. Journal of Solar Energy Engineering. 2012;**134**:034502-1

[9] Sabri N, Tlemçani A, Chouder A. Faults diagnosis in stand-alone photovoltaic system using artificial neural network. In: 2018 6th International Conference on Control Engineering & Information Technology (CEIT); 2018. pp. 1-6

[10] Sabri N, Tlemcani A, Chouder A. Intelligent fault supervisory system applied on stand-alone photovoltaic system. In: Proceedings of the 2018 International Conference on Applied Smart Systems (ICASS 2018); 2019

[11] Sabri N, Tlemçani A, Chouder A. Monitoring Tool for Stand-Alone Photovoltaic System Using Artificial Neural Network BT—Renewable Energy for Smart and Sustainable Cities. In: Hatti M, editor. Cham: Springer International Publishing; 2019. p. 114-21

[12] IEA-PVPS. A User Guide to Simple Monitoring and Sustainable Operation of PV-Diesel Hybrid Systems. Report IEA PVPS T9-16:2015. 2015

[13] Nailen RL. Battery protection-where do we stand? IEEE Transactions on Industry Applications. 1991;**27**(4): 658-667

[14] Purarjomandlangrudi A, Ghapanchi AH, Esmalifalak M. A data mining approach for fault diagnosis: An application of anomaly detection algorithm. Measurement [Internet]. 2014;**55**:343-352

[15] Zhao Y, Lehman B, Ball R, Mosesian J, de Palma J. Outlier detection rules for fault detection in solar photovoltaic arrays. In: 2013 Twenty-Eighth Annual IEEE Applied Power Electronics Conference and Exposition (APEC); 2013. pp. 2913-2920

[16] Zhao Y, Balboni F, Arnaud T, Mosesian J, Ball R, Lehman B. Fault experiments in a commercial-scale PV laboratory and fault detection using

local outlier factor. In: 2014 IEEE 40th Photovoltaic Specialist Conference (PVSC); 2014. pp. 3398-3403

[17] Castaner L, Silvestre S, Castaaner L. Modelling Photovoltaic Systems Using PSpice. Chichester: John Wiley & Sons, Ltd; 2002

[18] Villalva MG, Gazoli JR, Filho ER. Comprehensive approach to modeling and simulation of photovoltaic arrays. IEEE Transactions on Power Electronics. 2009;**24**(5):1198-1208

[19] Tremblay O, Dessaint LA. Experimental validation of a battery dynamic model for EV applications. World Electric Vehicle Journal. 2009;**03**: 0289-0298

[20] Marsland S. Novelty detection in learning systems. Neural Computing Surveys. 2002;**03**:1-39

[21] Markou M, Singh S. Novelty detection: A review—Part 1: Statistical approaches. Signal Processing [Internet]. 2003;**83**(12):2481-2497

[22] Chouder A, Silvestre S. Automatic supervision and fault detection of PV systems based on power losses analysis. Energy Conversion and Management. 2010;**51**(10):1929-1937